***FM 7-0** (FM 25-100)

Field Manual
No. 7-0

Headquarters
Department of the Army
Washington, DC, 22 October 2002

Training the Force

Contents

Preface ... iv

CHAPTER 1

How the Army Trains ... 1-1
The Training Imperative ... *1-1*
The Strategic Environment ... *1-2*
Joint, Interagency, Multinational (JIM) Training ... *1-4*
How the Army Trains the Army ... *1-4*
Leader Training and Development ... *1-6*
The Institutional Domain ... *1-7*
Initial Military Training (IMT) ... 1-8
Professional Military Education (PME) ... 1-8
The Operational Domain ... *1-9*
Commander's Responsibility ... 1-9
NCO Responsibility ... 1-9
Unit Responsibility ... 1-10
Relationship Between Institution and Unit ... 1-10
Operational Training and Major Exercises ... 1-10
The Self-Development Domain ... *1-11*
The Role of MACOMS, Corps, Divisions, USAR Regional Commands and ARNG Area Commands in Training ... *1-12*
Reserve Component Training ... *1-13*
Summary ... *1-14*

CHAPTER 2

Battle Focused Training ... 2-1
Principle of Training ... *2-1*
Commanders are Responsible for Training ... 2-2
NCOs Train Individuals, Crews, and Small Teams ... 2-2
Train as a Combined Arms and Joint Team ... 2-2
Train for Combat Proficiency ... 2-6
Train to Standard Using Appropriate Doctrine ... 2-6
Train to Adapt ... 2-7
Train to Maintain and Sustain ... 2-7
Train Using Multiechelon Techniques ... 2-7
Train to Sustain Proficiency ... 2-8
Train and Develop Leaders ... 2-9
Commanders and Training ... *2-10*

Distribution Restriction: Approved for public release; distribution is unlimited.

*This publication supersedes FM 25-100, 15 November 1988.

Develop and Communicate a Clear Vision ... 2-10
Train One Echelon Below and Evaluate Two Echelons Below ... 2-10
Require Subordinates to Understand and Perform Their Roles in Training ... 2-11
Train All Elements to be Proficient on Their Mission Essential Tasks ... 2-11
Develop Subordinates ... 2-12
Involve Themselves Personally in Planning, Preparing, Executing, and Assessing Training ... 2-12
Demand Training Standards are Achieved ... 2-12
Ensure Proper Task and Event Discipline ... 2-12
Foster a Command Climate That is Conducive to Good Training ... 2-12
Eliminate Training Distractions ... 2-13
Top-Down/Bottom-Up Approach to Training ... 2-13
Battle Focus ... 2-13
Army Training Management Cycle ... 2-14

CHAPTER 3

Mission Essential Task List Development ... 3-1
METL ... 3-2
METL Development Process ... 3-2
Inputs to METL Development ... 3-3
Wartime Operational Plans ... 3-3
Enduring Combat Capabilities ... 3-3
Operational Environment ... 3-3
Directed Missions ... 3-4
External Guidance ... 3-4
Commander's Analysis ... 3-5
Reserve Component METL Development ... 3-5
Echelon Above Division/Echelon Above Corps (EAD/EAC) METL Development ... 3-6
TDA METL Development ... 3-6
METL Development for Directed Missions ... 3-7
Joint METL (JMETL) Development ... 3-8
METL Development Fundamentals ... 3-8
METL Linked Training Strategy ... 3-10
Training Objectives ... 3-11
Battle Tasks ... 3-13

CHAPTER 4

Planning ... 4-1
Planning Process ... 4-1
Training Plans ... 4-4
Long-range Planning ... 4-6
Command Training Guidance (CTG) ... 4-8
Long-range Planning Calendar ... 4-9
Training and Time Management ... 4-9
Training Events ... 4-13
Live, Virtual, and Constructive (L-V-C) Training ... 4-15
Training Resources ... 4-17
Short-range Planning ... 4-22
Short-range Training Guidance ... 4-22
Short-range Planning Calendar ... 4-23
Training Events ... 4-24
Multiechelon Training ... 4-24
Training Resources ... 4-27

Train the Trainers 4-28
Short-range Training Briefings 4-28
Near-term Planning ***4-30***
Training Meetings 4-31
Training Schedules 4-31
CS and CSS Training ***4-32***
Garrison Training ***4-34***

CHAPTER 5

Execution **5-1**
Execution of Training ***5-1***
Preparation for Training ***5-2***
Conduct of Training ***5-3***
Recovery from Training ***5-6***
Role of Commanders and Senior Leaders ***5-7***
Role of Noncommissioned Officers ***5-7***

CHAPTER 6

Assessment **6-1**
Assessment ***6-1***
Organizational Assessment ***6-2***
Evaluations ***6-3***
Evaluation of Training ***6-4***
After Action Review ***6-4***
Evaluators ***6-5***
The Role of Senior Commanders and Leaders ***6-6***
Summary ***6-7***

Glossary **Glossary-1**
References **References-1**
Index **Index-1**

Preface

The U. S. Army exists for one reason—to serve the Nation. From the earliest days of its creation, the Army has embodied and defended the American way of life and its constitutional system of government. It will continue to answer the call to fight and win our Nation's wars, whenever and wherever they may occur. That is the Army's non-negotiable contract with the American people.

The Army will do whatever the Nation asks it to do, from decisively winning wars to promoting and keeping the peace. To this end, the Army must be strategically responsive and ready to be dominant at every point across the full spectrum of military operations.

Today, the Army must meet the challenge of a wider range of threats and a more complex set of operating environments while incorporating new and diverse technology. The Army meets these challenges through its core competencies: Shape the Security Environment, Prompt Response, Mobilize the Army, Forcible Entry Operations, Sustained Land Dominance and Support Civil Authorities. We must maintain combat readiness as our primary focus while transitioning to a more agile, versatile, lethal, and survivable Army.

Doctrine represents a professional army's collective thinking about how it intends to fight, train, equip, and modernize. When the first edition of FM 25-100, Training the Force, was published in 1988, it represented a revolution in the way the Army trains. The doctrine articulated by FMs 25-100, Training the Force, and 25-101, Battle Focused Training, has served the Army well. These enduring principles of training remain sound; much of the content of these manuals remains valid for both today and well into the future. FM 7-0 updates FM 25-100 to our current operational environment and will soon be followed by FM 7-1, which will update FM 25-101.

FM 7-0 is the Army's capstone training doctrine and is applicable to all units, at all levels, and in all components. While the examples in this manual are principally focused at division and below, FM 7-0 provides the essential fundamentals for all individual, leader, and unit training.

Training for warfighting is our number one priority in peace and in war. Warfighting readiness is derived from tactical and technical competence and confidence. Competence relates to the ability to fight our doctrine through tactical and technical execution. Confidence is the individual and collective belief that we can do all things better than the adversary and the unit possesses the trust and will to accomplish the mission.

FM 7-0 provides the training and leader development methodology that forms the foundation for developing competent and confident soldiers and units that will win decisively in any environment. Training is the means to achieve tactical and technical competence for specific tasks, conditions, and standards. Leader Development is the deliberate, continuous, sequential, and progressive process, based on Army values, that develops soldiers and civilians into competent and confident leaders capable of decisive action.

Closing the gap between training, leader development, and battlefield performance has always been the critical challenge for any army. Overcoming this challenge requires achieving the correct balance between training management and training execution. Training management focuses leaders on the science of training in terms of resource efficiencies (such as people, time, and ammunition) measured against tasks and standards. Training execution focuses leaders on the art of leadership to develop trust, will, and teamwork under varying conditions—intangibles that must be developed to win decisively in combat. Leaders integrate this science and art to identify the right tasks, conditions, and standards in training, foster unit will and spirit, and then adapt to the battlefield to win decisively.

FM 7-0 provides the Training Management Cycle and the necessary guidelines on how to plan, execute, and assess training and leader development. Understanding "How the Army Trains the Army" to fight is key to successful joint, interagency, multinational (JIM), and combined arms operations. Effective training leads to units that execute the Army's core competencies and capabilities.

All leaders are trainers! This manual is designed for leaders at every level and in every type of organization in the Army.

The proponent for this publication is U.S. Army Training and Doctrine Command (TRADOC). Send comments and recommendations on DA Form 2028 to Commander, HQ TRADOC, ATTN: ATTG-ZA, Fort Monroe, Virginia 23651-5000.

Direct e-mail questions to the following address: doctrine@Monroe.army.mil.

Unless this publication states otherwise, masculine nouns and pronouns do not refer exclusively to men.

Chapter 1

How the Army Trains

- *Develop trust soldier-to-soldier, leader to led, unit-to-unit in the Army and grow the warfighting confidence of the force.*
- *Train for decisive warfighting.*
- *Train soldiers now, and grow leaders for the next conflict.*
- *Ensure that our soldiers are physically and mentally prepared to dominate the next battlefield—no soldier goes into harm's way untrained.*
- *Our soldiers must be comfortable and confident in the elements—fieldcraft, fieldcraft, fieldcraft.*

General Eric Shinseki

THE TRAINING IMPERATIVE

1-1. Every soldier, noncommissioned officer (NCO), warrant officer, and officer has one primary mission—to be trained and ready to fight and win our Nation's wars. Success in battle does not happen by accident; it is a direct result of tough, realistic, and challenging training. The Army exists to deter war, or if deterrence fails, to reestablish peace through victory in combat wherever U.S. interests are challenged. To accomplish this, the Army's forces must be able to perform their assigned strategic, operational, and tactical missions. For deterrence to be effective, potential enemies must know with certainty that the Army has the credible, demonstrable capability to mobilize, deploy, fight, sustain, and win any conflict. Training is the process that melds human and materiel resources into these required capabilities. The Army has an obligation to the American people to ensure its soldiers go into battle with the assurance of success and survival. This is an obligation that only rigorous and realistic training, conducted to standard, can fulfill.

1-2. We train the way we fight because our historical experiences show the direct correlation between realistic training and success on the battlefield. Today's leaders must apply the lessons of history in planning training for tomorrow's battles. We can trace the connection between training and success in battle to our Army's earliest experiences during the American Revolution. General Washington had long sensed the need for uniform training and organization and, during the winter of 1777-1778 while camped at Valley Forge, he secured the appointment of Von Steuben, a Prussian, as inspector general in charge of training. Von Steuben clearly understood the difference between the American citizen-soldier and the European professional. He noted early that American soldiers had to be told why they did things before they would do them well, and he applied this philosophy in his training. It helped the Continental soldiers understand and endure the rigorous and demanding training he put them through. After Valley Forge, Continentals would fight on equal terms with British Regulars. Von Steuben began the tradition of effective unit level training that today still develops leaders and forges battle-ready units for the Army.

1-3. Over two centuries later, the correlation between tough, realistic training and success on the battlefield remains the same. During Operation Enduring Freedom, and Operation Anaconda in Afghanistan, the U.S. Army deployed a trained and ready force on short notice to a contemporary battlefield fighting against a coalition of rebel forces on difficult terrain.

1-4. These units trained to their wartime mission, and developed company grade officers, NCOs, and soldiers who knew their jobs and were confident they could act boldly and decisively. Their confidence, and technical and tactical competence gave them the ability to adapt to the mission and harsh environment with resounding success. Airmobile infantry quickly perfected methods of routing rebel forces from heavily fortified caves. Special forces teams rode horses with their host nation counterparts—learning to call in tactical air support with devastating accuracy while on the move. Staffs quickly learned how to integrate Special Operations Forces (SOF) and conventional force operations. Engineer units cleared mine fields that were as old as many of their soldiers involved in the clearing process. Again, American soldiers had met the enemy and decisively defeated them.

1-5. The Army's battle-focused training was validated. These soldiers trained as they planned to fight and won. Their success was due to the Army's emphasis on battle focused training which emphasized training essential warfighting tasks to standard and building cohesive combined arms teams able to adapt to the mission. Army units today train, alert, and deploy prepared for combat. Their battle focused training experience gives them the flexibility to continue training and adapting to the mission as it evolves.

THE STRATEGIC ENVIRONMENT

1-6. In an era of complex national security requirements, the Army's strategic responsibilities now embrace a wider range of missions that present even greater challenges in our training environment. To "train the way we fight," commanders and leaders at all levels must conduct training with respect to a wide variety of operational missions across the full spectrum of operations. These operations may include combined arms, joint, multinational, and interagency considerations, and span the entire breadth of terrain and environmental possibilities. Commanders must strive to set the daily training conditions as closely as possible to those expected for actual operations.

1-7. The operational missions of the Army include not only war, but also military operations other than war (MOOTW). Operations may be conducted as major combat operations, a small-scale contingency, or a peacetime military engagement. Offensive and defensive operations normally dominate military operations in war along with some small-scale contingencies. Stability operations and support operations dominate in MOOTW. Commanders at all echelons may combine different types of operations simultaneously and sequentially to accomplish missions in war and MOOTW. Throughout this document, we will emphasize the primary function of the Army—to fight and win our Nation's wars. Implicit in the emphasis is the mounting importance of MOOTW. These missions also require training; future conflict will likely involve a mix of combat and MOOTW, often concurrently. The range of possible missions complicates training. Army forces cannot train for every possible mission; they train for war and prepare for specific missions as time and circumstances permit. The nature of world crises requires Army forces to simultaneously train, deploy, and execute. Therefore, at Army level, warfighting will encompass the full spectrum of operations that the Army may be called upon to execute. Warfighting in units is refined and focused on assigned wartime missions or directed change of missions.

1-8. Contingency operations in the 1990s normally followed a sequence of alert, train, deployment, extended build-up, and shaping operations followed by a period of decisive operations. To be truly responsive and meet our commitments, Army forces must be deployable and capable of rapidly concentrating combat power in an operational area with minimal additional training. Our forces today use a train, alert, deploy sequence. We cannot count on the time or opportunity to correct or make up training deficiencies after deployment. Maintaining forces that are ready now, places increased emphasis on training and the priority of training. This concept is a key link between operational and training doctrine.

1-9. Units train to be ready for war based on the requirements of a precise and specific mission; in the process, they develop a foundation of combat skills, which can be refined based on the requirements of the assigned mission. Upon alert, commanders assess and refine from this foundation of skills. In the train, alert, deploy process commanders use whatever time the alert cycle provides to continue to refine mission-focused training. Training continues during time available between alert notification and deployment, between deployment and employment, and even during employment as units adapt to the specific battlefield environment and assimilate combat replacements.

1-10. Resources for training are not unconstrained and compete with other missions and activities. Time is the inelastic resource, there is not enough and it cannot be increased. We cannot do everything; we must forge and sustain trained and ready forces. Training for the warfight, training to maintain near-term readiness is the priority; compliance training and non-mission activities are of lower priority. If training cannot be conducted, readiness reports are the vehicle to inform the Army's leadership of the risks being assumed.

1-11. The key to winning on the battlefield is the understanding of "how we fight" and the demonstrated confidence, competence, and initiative of our soldiers and leaders. Training is the means to achieve the tactical and technical proficiency that soldiers, leaders, and units must have to enable them to accomplish their missions. Training focuses on fighting and winning battles. The proficiency derived from this training is the same required for many MOOTW tasks. The ability to integrate and synchronize all available assets to defeat any enemy tactically gives our Army great credibility and respect that enhances our ability to accomplish all missions to include MOOTW.

1-12. Responsibility for success on the future battlefield rests on the shoulders of today's Army leaders at all levels. To ensure this success, all leaders must focus training on warfighting skills, and make that training the priority.

JOINT, INTERAGENCY, MULTINATIONAL (JIM) TRAINING

1-13. The purpose of joint training is to prepare the Army to execute missions as a part of a joint force in the conduct of joint military operations and across the full spectrum of conflict. Employing Army forces at the right place and time allows combatant commanders to conduct decisive land operations along with air, sea, and space-based operations. The Army provides to a joint force commander (JFC) trained and ready forces that expand the commander's range of military options. Army commanders tailor and train forces to react quickly to any crisis.

1-14. Commanders of major Army headquarters may serve as the joint force land component commander (JFLCC), a combined forces commander (CFC), or as the joint task force commander (JTFC). To perform these assignments organizations conduct joint training.

1-15. Joint training uses joint doctrine, tactics, techniques, and procedures, and the training involves more than one Service component. However, two or more Services training together using their respective service doctrine, tactics, techniques, and procedures are Service-sponsored interoperability training. Although, not classified as joint training, Service sponsored interoperability is a vital component of joint proficiency and readiness.

1-16. Multinational training is based on applicable multinational, joint and/or service doctrine and is designed to prepare organizations for combined operations with allied nations.

1-17. Interagency training is based on applicable standard operating procedures; and, is designed to prepare the Army to operate in conjunction with government agencies.

1-18. The Army training doctrine contained in this manual provides Army commanders the tools to develop experienced leaders and adaptive organizations prepared to exercise command and control of joint and multinational forces, and to provide interagency unity of effort.

HOW THE ARMY TRAINS THE ARMY

1-19. Training is a team effort and the entire Army—Department of the Army, major Army commands (MACOMs), the institutional training base, units, the combat training centers (CTC), each individual soldier and the civilian work force—has a role that contributes to force readiness. Department of the Army and MACOMs are responsible for resourcing the Army to train. The institutional Army including schools, training centers, and NCO academies, for example, train soldiers and leaders to take their place in units in the Army by teaching the doctrine and tactics, techniques, and procedures (TTP). Units, leaders, and individuals train to standard on their assigned missions, first as an organic unit and then as an integrated component of a team. Operational deployments, and major training opportunities such as major training exercises, CTCs, CTC-like training, and external evaluations (EXEVAL) provide rigorous, realistic, and stressful training and operational experience under actual or simulated combat and operational conditions to enhance unit readiness and produce bold, innovative leaders. Simultaneously, individual soldiers, NCOs, warrant officers, officers, and the civilian work force are responsible for training themselves through personal self-development. Training is a continuous, lifelong endeavor that produces competent, confident, disciplined, and adaptive soldiers and leaders with the warrior ethos in our Army. Commanders have the ultimate responsibility to train soldiers and develop leaders who can adjust to change with confidence and exploit new situations, technology, and developments to their advantage. The result of this Army-wide team effort is a training and leader development system that is unrivaled in the world. Effective training produces the force—soldiers, leaders, and units—that can successfully execute any assigned mission.

1-20. The Army Training and Leader Development Model (figure 1-1) centers on developing trained and ready units led by competent and confident leaders. The model identifies an important interaction that trains soldiers now and develops leaders for the future. Leader Development is a lifelong learning process. The three core domains that shape the critical learning experiences throughout a soldier's and leader's career are the operational, institutional, and self-development domains. Together, these domains interact using feedback and assessment from various sources and methods to maximize warfighting readiness. Each domain has specific, measurable actions that must occur to develop our leaders. The operational domain includes home station training, combat training center rotations, joint training exercises, and operational deployments that satisfy national objectives. Each of these actions provides foundational experiences for soldiers, leaders, and unit development. The institutional domain focuses on educating and training soldiers and leaders on the key knowledge, skills, and attributes required to operate in any environment. It includes individual, unit and joint schools, and advanced education. The self-development domain, both structured and informal, focuses on taking those actions necessary to reduce or eliminate the gap between operational and institutional experiences. Throughout this lifelong learning and experience process, there is formal and informal assessment and feedback of performance to prepare leaders for their next level of responsibility. Assessment is the method used to determine the proficiency and potential of leaders against a known standard. Feedback must be clear, formative guidance directly related to the outcome of training events measured against standards.

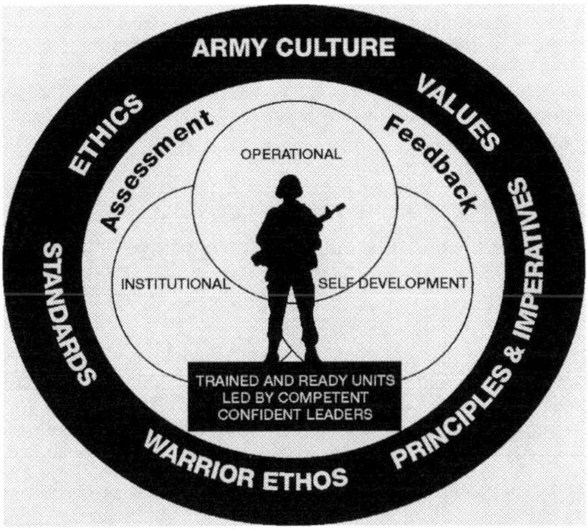

Figure 1-1. Army Training and Leader Development Model

1-21. The importance of training the technical skills to develop competent soldiers and leaders must be directly linked to creating confident soldiers, leaders, and units with the will and warrior spirit to dominate in any environment. The operational, institutional, and self-development domains are influenced by and adapted based on the overall strategic context of the Army. Joint, interagency, and multinational training, education, and individual assignment experiences shape the competence and confidence of leaders and units.

1-22. All of these interrelated activities take place within the Army's culture or shared set of beliefs, values, and assumptions that define for us what is most important. Our culture is ingrained in our new soldiers and reinforced daily to all of us in order to provide a positive framework for everything we do. A detailed discussion of Army culture will be addressed in FM 6-22, Leadership, and the updated version of DA PAM 350-58, Leader Development for America's Army.

LEADER TRAINING AND DEVELOPMENT

1-23. The Army is a profession, the Profession of Arms. Warfighting in defense of U. S. values and interests is the core competency of this profession. As a profession, the development of each member becomes the foundation, involving a lifelong devotion to duty both while in uniform and upon return to the civilian life. Professional development involves more than mastering technical skills. What is uniquely distinct to the military profession is its emphasis on not only what is to be accomplished, but how it is accomplished and with the full realization that the profession of arms may require of its members, the supreme sacrifice. Professional development extends to inculcating the Army values of Loyalty, Duty, Respect, Selfless Service, Integrity, Honor, and Personal Courage in every soldier to create a warrior ethos based on camaraderie and service to our Nation. Professional education provides the foundation involving a variety of training domains ranging from institutional schooling, self-study, and operational experience to personal interaction with superiors, peers, and subordinates. All of these interactions are essential in developing and understanding training and leader development for warfighting.

1-24. Competent and confident leaders are a prerequisite to the successful training of ready units. It is important to understand that leader training and leader development are integral parts of unit readiness. Leaders are inherently soldiers first and should be technically and tactically proficient in basic soldier skills. They are also adaptive, capable of sensing their environment, adjusting the plan when appropriate, and properly applying the proficiency acquired through training.

1-25. Leader training is an expansion of these skills that qualifies them to lead other soldiers. As such, the doctrine and principles of training leader tasks is the same as that for any other task set forth in FM 7-0 and requires the same level of attention of senior commanders. Leader training occurs in the institutional Army, the unit, the combat training centers, and through self-development. Leader training is just one portion of leader development.

1-26. Leader development is the deliberate, continuous, sequential, and progressive process, grounded in Army values, that grows soldiers and civilians into competent and confident leaders capable of decisive action. Leader development is achieved through the lifelong synthesis of the knowledge, skills, and experiences gained through institutional training and education, organizational training, operational experience, and self-development. Commanders play the key role in leader development that ideally produces tactically and technically competent, confident, and adaptive leaders who act with boldness and initiative in dynamic, complex situations to execute mission-type orders achieving the commander's intent.

THE INSTITUTIONAL DOMAIN

1-27. The institutional Army (schools and training centers) is the foundation for lifelong learning. The institution is a key enabler for unit readiness. It develops competent, confident, disciplined, and adaptive leaders and soldiers able to succeed in situations of great uncertainty. The institution provides the framework to develop future leadership characteristics that produce critical thinkers capable of full spectrum visualization, systems understanding, and mental agility. Institutional training and education enhances military knowledge, individual potential, initiative, and competence in warfighting skills. It infuses an ethos of service to the Nation and the Army, and provides the educational, intellectual, and experiential foundation for success on the battlefield. The institution teaches Army doctrine and provides the experiences that train leaders and soldiers. It trains them to adapt to uncertainty and be creative and innovative problem solvers as members of lethal units and battle staffs in combined arms, and JIM operations. Institutions provide training on common tasks and a selected portion of occupation-related critical tasks, and continue to provide lifelong, through mutual reach, access to training materials for individual soldier or unit use. The elements of institutional training and education include—

INITIAL MILITARY TRAINING (IMT)

1-28. This training provides the basic skills, knowledge, and task proficiency to become a soldier and subsequently to succeed as members of a small Army unit, contribute to unit mission accomplishment, and survive on the battlefield. IMT is the foundation training given to all personnel upon entering the Army. It provides an ordered transition from being a civilian to becoming a soldier, motivation to become a dedicated and productive member of the Army, and qualification on basic critical soldier skills and knowledge. IMT instills an appreciation for the Army in a democratic society, inspires the Army warrior ethos, and establishes Army values of loyalty, duty, respect, selfless service, honor, integrity, and personal courage. Newly commissioned officers will be competent and confident small unit leaders trained in fieldcraft. Warrant officers will be technically proficient in the systems associated with their functional specialty. Enlisted soldiers will be qualified in the critical military occupational specialty tasks and standards defined by their branch proponent. The soldierization and professional development process continues under the leadership of NCOs when these new soldiers arrive in their first unit.

PROFESSIONAL MILITARY EDUCATION (PME)

1-29. PME develops Army leaders. Officer, warrant officer, and NCO training and education is a continuous, career-long, learning process that integrates structured programs of instruction—resident at the institution and non-resident via distributed learning at home station. PME is progressive and sequential, provides a doctrinal foundation, and builds on previous training, education and operational experiences. PME provides hands-on technical, tactical, and leader training focused to ensure leaders are prepared for success in their next assignment and higher-level responsibility.

- Officer Education System (OES). Army officers must lead and fight; be tactically and technically competent; possess leader skills; understand how the Army operates as a service, as well as a component of a joint, multinational, or interagency organization; demonstrate confidence, integrity, critical judgment, and responsibility; operate in a complex, uncertain, and rapidly changing environment; build effective teams amid continuous organizational and technological change; and solve problems creatively. OES develops officers who are self-aware and adaptive to lead Army units to mission success.
- Warrant Officer Education System (WOES). Warrant officers are the Army's technical experts. WOES develops a corps of highly specialized experts and trainers who are fully competent and proficient operators, maintainers, administrators, and managers of the Army's equipment, support activities, and technical systems.

- NCO Education System (NCOES). NCOES trains NCOs to lead and train soldiers, crews, and subordinate leaders who work and fight under their leadership. NCOES provides hands-on technical, tactical, and leader training focused to ensure that NCOs are prepared for success in their next assignment and higher-level responsibility.
- Functional Training. In addition to the preceding PME courses, there are functional courses available in both resident and non-resident distributed learning modes that enhance functional skills for specific duty positions. Examples are Battalion S2, Battalion Motor Officer, First Sergeant, Battle Staff NCO, and Airborne courses.

THE OPERATIONAL DOMAIN

1-30. Soldier and leader training and development continue in the unit. Using the institutional foundation, training in organizations and units focuses and hones individual and team skills and knowledge.

COMMANDER'S RESPONSIBILITY

1-31. The unit commander is responsible for the wartime readiness of all elements in the formation. The commander is, therefore, the primary trainer of the organization, responsible for ensuring that all training is conducted in accordance with the unit's mission essential task list (METL) to the Army standard. This is the commander's number one priority. The command climate must reflect this priority. The commander analyzes the unit's wartime mission and develops the unit's METL. Using appropriate doctrine and mission training plans (MTPs), the commander plans training and briefs the training plan to the senior commander. The senior commander is responsible for resourcing, ensuring stability and predictability, protecting training from interference, and executing and assessing training. Commanders ensure MTP standards are met during all training. If they are not, the unit must retrain until the tasks are performed to standard. Train to standard, not to time.

1-32. Key to effective unit training is the commander's involvement and presence in planning, preparing, executing, and assessing unit training to standard. Commanders ensure MTP standards are met during all training. If a squad, platoon, or company fails to meet established standards for identified METL tasks, the unit must retrain until the tasks are performed to standard. Training to standard on METL tasks is more important than completion of an event such as an EXEVAL. Focus on sustaining METL proficiency—this is the critical factor commanders must adhere to when training small units.

NCO RESPONSIBILITY

1-33. A great strength of the U.S. Army is its professional NCO Corps who take pride in being responsible for the individual training of soldiers, crews, and small teams. They ensure the continuation of the soldierization process of new soldiers when they arrive in the unit. Within the unit, the NCO support channel (leadership chain) parallels and complements the chain of command. It is a channel of communication and supervision from the command sergeant major (CSM) to first sergeant and then to other NCOs and enlisted personnel. In addition, NCOs train soldiers to the non-negotiable standards published in MTPs and soldiers training publications (STP). Commanders will define responsibilities and authority of their NCOs to their staffs and subordinates.

UNIT RESPONSIBILITY

1-34. Unit training consists of three components: collective training that is derived directly from METL and MTPs, leader development that is embedded in the collective training tasks and in discrete individual leader focused training, and individual training that establishes, improves, and sustains individual soldier proficiency in tasks directly related to the unit METL. Commanders conduct unit training to prepare soldiers and leaders for unit missions. All units concentrate on improving and sustaining unit task proficiency.

RELATIONSHIP BETWEEN INSTITUTION AND UNIT

1-35. The goal of unit training is to develop and sustain the capability to deploy rapidly, and to fight and win as part of a combined arms team in a variety of operational and organizational environments. Training in both the institution and the unit works together toward achieving this goal. Institutions provide foundational training and education and, when combined with individual unit experience, provide soldiers and leaders what they need to succeed in each subsequent level of service throughout their careers, appropriate to new and increasing levels of responsibility. The institutions also provide reach-back capability for functional and duty position-related training or reference materials throughout a soldier's service. Unit commanders, through subordinate leaders, build on the foundation provided by Army schools to continue developing the skills and knowledge required for mission success, as articulated in the unit's METL. Unit commanders are responsible for sustaining small unit leader and individual soldier skills to support the unit's mission. Institutions are responsible to stay abreast of requirements and developments in the field to ensure the foundations they set prepare soldiers for duty in their units.

OPERATIONAL TRAINING AND MAJOR EXERCISES

1-36. Leader, individual soldier, and unit training and development continue during operational missions and major training events. These events enhance leader development and combat readiness. They improve leader skills and judgment while increasing unit collective proficiency through realistic and challenging training and real-time operational missions.

1-37. Major training events such as situational training exercises (STX), EXEVALs, and deployment exercises provide feedback to assist commanders in assessing the effectiveness of their leader, individual soldier, unit, and maintenance training programs. Units and individuals establish and sustain their tactical and technical training proficiency. Leaders learn to solve tactical problems, and to give appropriate and meaningful orders. They get feedback on the quality of their decisions and obtain an understanding of impact that the frictions of the battlefield have on their decisions. Adaptive leaders are tactically and technically competent, confident in their abilities, and routinely demonstrate initiative within the framework of their commander's intent. Major training events provide experiences that contribute to developing leader, soldier, and unit adaptiveness.

1-38. The CTC Program, consisting of the National Training Center (NTC), Joint Readiness Training Center (JRTC), Combat Maneuver Training Center (CMTC), Battle Command Training Program (BCTP), and other CTC-like training provides highly realistic and stressful joint and combined arms training based on current doctrine. Commanders fight with the equipment they would expect to take to war, arrayed against a free-thinking, opportunistic opposing force (OPFOR) with an equal chance to win, monitored by a dedicated, well-trained, and experienced observer/controller team. Consequences of tactical decisions are fully played out in scenarios where the outcome is not assured. Doctrine-based after action reviews (AAR) identify strengths and shortcomings in unit planning, preparation, and execution, and guide leaders to accept responsibility for shortcomings and produce a fix. The CTC Program is the Army's premier training and leader development experience. It provides the following tangible benefits to the Army:

- Produces bold, innovative leaders through stressful tactical and operational exercises.
- Embeds doctrine throughout the Army.
- Provides feedback to assist the commander in assessing unit readiness.
- Provides feedback to Army, JIM participants.
- Provides a deployable capability to export observer/controllers, instrumentation, and the AAR process to units at locations other than a CTC.
- Provides a data source for lessons learned and trends to improve doctrine, training, leader development, organization, materiel, and soldier considerations.

1-39. Operational missions—whether they are combat operations, such as in Afghanistan, or stability operations, such as in Bosnia—continue training and leader development. Operational missions validate the fundamentals of leadership, planning, and training. Unit and individual proficiency is evaluated, and leaders are trained and developed. AARs are conducted, strengths are maintained, and weaknesses are corrected. These missions provide significant experience for our leaders, soldiers, and units. The experiences from these missions feed back to the institution to support doctrine development, and other leader, soldier, and unit training.

THE SELF-DEVELOPMENT DOMAIN

1-40. Learning is a lifelong process. Institutional, organizational, and operational training alone cannot provide the insight, intuition, imagination, and judgment needed in combat. The gravity of our profession requires comprehensive self-study and training. In no other profession is the cost of being unprepared so high. Soldiers and leaders at all levels continually study our profession in preparation to fight and win our Nation's wars. This requires commanders at all levels to create an environment that encourages subordinates to establish personal and professional development goals. Further refinement of those interests should occur through personal mentoring by commanders and first line leaders. Application of battle-focused officer and NCO professional development programs are essential to leader development. Exploiting reach-back, distributed learning, and continuing education technologies support these programs.

1-41. Self-development is continuous and should be emphasized in both institutional and operational assignments. Successful self-development requires a team effort. Self-development starts with an assessment of individual strengths, weaknesses, potential, and developmental needs. Commanders and leaders provide feedback to enable subordinates to determine the reasons for their strengths and weaknesses. Together, they prioritize self-development goals and determine courses of action to improve performance. Self-development is—

- A planned process involving the leader and the subordinate being developed. It enhances previously acquired skills, knowledge, behaviors, and experience; contributes to personal development; and highlights the potential for progressively more complex and higher-level assignments. Self-development focuses on maximizing individual strengths, minimizing weaknesses, and achieving individual development goals.
- Initial self-development is very structured and generally narrow in focus. The focus broadens as individuals understand their strengths and weaknesses, determine their individual needs, and become more experienced. Each soldier's knowledge and perspective increases with experience, institutional training, and operational assignments. It is accelerated and broadened by specific, goal-oriented self-development actions.

THE ROLE OF MACOMS, CORPS, DIVISIONS, USAR REGIONAL COMMANDS AND ARNG AREA COMMANDS IN TRAINING

1-42. These commands, whether oriented along operational, functional, or specialty missions, have unique responsibilities for managing and supporting training. Their most important contribution to training is to establish stability in the training environment by maintaining focus on warfighting tasks, identifying and providing resources, protecting planned training, and providing feedback that produces good training and develops good trainers and leaders.

1-43. The corps' and divisions' fundamental basis for organization and operations is combined arms operations. They conduct these operations increasingly in JIM environments. Corps commanders' training focus is on warfighting, to include joint operations, and training division commanders and corps separate commands and brigades.

1-44. Corps and division commanders must integrate SOF into their training plans. This provides opportunities to explore new combinations of concepts, people, organizations, and technology that expand their capabilities and enhance interoperability and leverage other service capabilities.

1-45. Warfighting is the corps' and division's top priority. Corps and division commanders have a pivotal role in the Army Training Management System as the guidance and decisions they provide brigade and battalion commanders directly affect the planning and execution of training at the company level.

1-46. The MACOMs, corps, and divisions ensure that competencies are trained to standard. When commanders do this they make their greatest contribution to leader development and unit readiness.

RESERVE COMPONENT TRAINING

1-47. The Army consists of the active component (AC) and the Reserve Components (RC). The AC is a federal force of full-time soldiers and Department of the Army civilians. The RC consists of the ARNG, the USAR, and their civilian support personnel. Each component is established under different statutes and has unique and discrete characteristics, but all share the same doctrine and training process, and train to the same standard. Availability of training support system (TSS) capabilities, however, varies between components. All train to the same standard; however, the RC trains at lower echelons. The number of tasks trained will usually differ as a result of the training time available; the conditions may vary based on the RC unique environment.

1-48. The RC represent a large portion of the Army's deterrence and warfighting power. They are an integral part of the force. However, available training time has a significant impact on RC training. RC units have a limited number of available training days. Geographic dispersion of units also impacts RC training. An average reserve battalion is spread over a 150- to 300-mile radius. Additionally, most reserve units travel an average of 150 miles to the nearest training area. Individual soldiers often travel an average of 40 miles to their training sites.

1-49. RC units recruit many of their own soldiers. Since these new recruits may be assigned to the RC unit prior to completion of IMT, the RC may have fewer military occupational specialty (MOS) qualified personnel assigned than their AC counterparts. Additionally, even though doctrine requires trained leaders to train units and soldiers, RC leaders may be unable to attend professional military education until after assigned to their units. Priority of training for RC units will go to individual duty military occupational specialty qualification (DMOSQ) and professional development to produce qualified soldiers and leaders.

1-50. RC units have premobilization readiness and postmobilization training requirements. Premobilization readiness plans must be developed and approved for the current fiscal and training year. Similarly, postmobilization plans must be developed and approved for units with deployment missions. For example, the RC focuses premobilization training for infantry, armor and cavalry units on platoon and lower level maneuver and collective tasks and drills. Postmobilization training focuses on platoon gunnery, company team, and higher-level collective tasks. IMT and professional military education requirements for individual reserve officers and soldiers approximate that of the active Army with training provided by the institution. In sum, RC units focus on fewer tasks done to standard during premobilization training.

SUMMARY

1-51. Army training has one purpose—to produce competent, confident, adaptive soldiers, leaders and units, trained and ready to fight and win our Nation's battles. The Army training and leader development model integrates institutional, operational, and individual self-development into a training management system. The commander is responsible for unit training and integrates the institutional, operational, as well as individual self-development resources to train combat ready units. Commanders are responsible for the wartime readiness of every aspect of their unit, while NCOs train individual soldiers, crews, and teams. All training focuses on the METL and all factors involved in training lead to unit readiness. Training is the Army's number one priority. Training is WHAT we do, not SOMETHING we do.

Chapter 2

Battle Focused Training

The key to fighting and winning is an understanding of "how we train to fight" at every echelon. Training programs must result in demonstrated tactical and technical competence, confidence, and initiative in our soldiers and their leaders. Training will remain the Army's top priority because it is the cornerstone of combat readiness!

General Carl E. Vuono

Commanders train their units to be combat ready. Training is their number one priority. Commanders achieve this using tough, realistic, and challenging training. At every echelon, commanders must train their unit to the Army standard. Battle focus enables the commander to train units for success on the battlefield. Using the Army Training Management Cycle, the commander continuously plans, executes, and assesses the state of training in the unit. This cycle provides the framework for commanders to develop their unit's METL, establish training priorities, and allocate resources.

Commanders and leaders at all echelons use the Principles of Training discussed in this chapter to develop and execute effective training. As commanders train their units on METL tasks, senior commanders reinforce training by approving and protecting training priorities and providing resources.

PRINCIPLE OF TRAINING

2-1. There are 10 Principles of Training.

- **Commanders are responsible for training.**
- **NCOs train individuals, crews, and small teams.**
- **Train as a combined arms and joint team.**
- **Train for combat proficiency.**
  - **Realistic conditions.**
  - **Performance-oriented.**
- **Train to standard using appropriate doctrine.**
- **Train to adapt.**
- **Train to maintain and sustain.**
- **Train using multiechelon techniques.**
- **Train to sustain proficiency.**
- **Train and develop leaders.**

Figure 2-1. Principles of Training

COMMANDERS ARE RESPONSIBLE FOR TRAINING

2-2. Commanders are responsible for the training and performance of their soldiers and units. They are the primary training managers and trainers for their organization, are actively engaged in the training process, and adhere to the 10 principles of training in figure 2-1. To accomplish their training responsibility, commanders must—

- Be present at training to maximum extent possible.
- Base training on mission requirements.
- Train to applicable Army standards.
- Assess current levels of proficiency.
- Provide the required resources.
- Develop and execute training plans that result in proficient individuals, leaders, and units.

2-3. Commanders delegate authority to NCOs in the support channel as the primary trainers of individuals, crews, and small teams. Commanders hold NCOs responsible for conducting standards-based, performance-oriented, battle-focused training and provide feedback on individual, crew, and team proficiency.

NCOS TRAIN INDIVIDUALS, CREWS, AND SMALL TEAMS

2-4. NCOs continue the soldierization process of newly assigned enlisted soldiers, and begin their professional development. NCOs are responsible for conducting standards-based, performance-oriented, battle-focused training. They–

- Identify specific individual, crew, and small team tasks that support the unit's collective mission essential tasks.
- Plan, prepare, rehearse, and execute training.
- Evaluate training and conduct AARs to provide feedback to the commander on individual, crew, and small team proficiency.

2-5. Senior NCOs coach junior NCOs to master a wide range of individual tasks.

TRAIN AS A COMBINED ARMS AND JOINT TEAM

2-6. The Army provides a JFC with trained and ready forces that expand the command's range of military options in full spectrum operations. Army commanders tailor and train forces to react quickly to any crisis. Army forces provide a JFC the capability to—

- Seize areas previously denied by the enemy.
- Dominate land operations.
- Provide support to civil authorities.

2-7. Army forces seldom operate unilaterally. Joint interdependence from the individual, crew, and small team to the operational level requires training to develop experienced, adaptive leaders, soldiers, and organizations prepared to operate with joint, and multinational forces and to provide interagency unity of effort.

2-8. The fundamental basis for the organization and operation of Army forces is combined arms. Combined arms is the integrated application of several arms to achieve an effect on the enemy that is greater than if each arm was used against the enemy separately or in sequence. Integration involves arrangement of battlefield actions in time, space, and purpose to produce maximum relative effects of combat power at a decisive place and time. Through force tailored organizations, commanders and their staffs integrate and synchronize the battlefield operating systems (BOS) to achieve combined arms effects and accomplish the mission.

2-9. Today's Army doctrine requires teamwork at all echelons. Well-trained Army combined arms teams can readily perform in JIM environments. When committed to battle, each unit must be prepared to execute operations without additional training or lengthy adjustment periods. Leaders must regularly practice of habitually associated combat arms, combat support, and combat service support capabilities. Teams can only achieve combined arms proficiency and cohesiveness when they train together. Similarly, peacetime relationships must mirror wartime task organization to the greatest extent possible.

2-10. Commanders are responsible for training all warfighting systems. The full integration of the combined arms team is attained through the task organization approach to training management. Task organizing is a temporary grouping of forces designed to accomplish a particular mission. This approach acknowledges that the maneuver commander integrates and synchronizes the BOS. In short, the maneuver commander, assisted by higher echelon leaders, forges the combined arms team. An example of a *task-organized brigade* and its warfighting systems is depicted at figure 2-2.

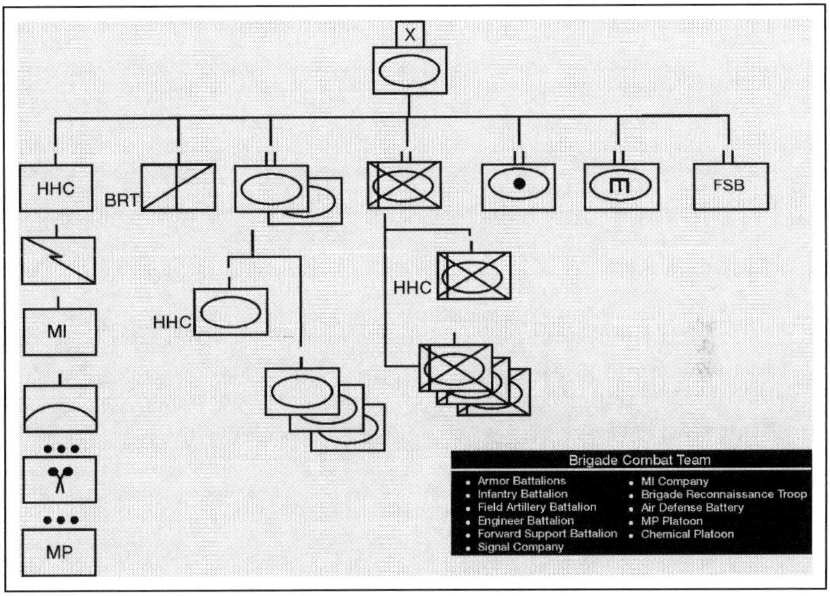

Figure 2-2. Brigade Combat Team

2-11. The commander of the *task-organized* force must develop a training plan that addresses two complementary challenges. The commander's training plan must achieve combined arms proficiency and ensure functional training proficiency of the combat arms, combat support, and combat service support units of the task force. Combined arms proficiency requires effective integration of BOS functions. Effective integration of BOS results in synchronization. Functional BOS proficiency is fundamental for effective BOS integration. The commander's training plan must integrate combined arms and functional training events.

2-12. Combined arms training is standards based. The independent training of functional tasks and combined arms tasks to standard will not guarantee the desired effects of applying combat power at a decisive place and time. The standard for effective combined arms training requires a sequenced and continuous execution of functional tasks and combined arms tasks to standard in order to achieve "...integrated relative combat power at a decisive place and time."

2-13. The role of commanders and NCOs in combined arms training cannot be overemphasized. Commanders have training responsibilities that encompass both BOS functional task proficiency and special staff officer combined arms task proficiency. Likewise, NCOs have similar training responsibilities to ensure BOS related individual and crew functional task proficiency, as well as, individual and staff section related combined arms task proficiency. Combined arms training requires commanders' and NCOs' active involvement in all phases of training.

2-14. Functional proficiency requires expertise in a particular BOS function, its capabilities, and its requirements. Organizations that provide elements of a specific BOS function, such as corps support command and divisional air defense artillery battalion, must train to maintain their functional proficiency. Integration involves expertise in coordination among functional troop unit commanders and staffs, and other functional commanders and staffs.

2-15. The combined arms training challenge is the same for all echelons of command. The complexity, however, increases at each higher echelon of command. The tempo, scope, and scale of operations at higher command echelons increase coordination requirements for planning and executing staff, joint, multinational, and interagency training. Commanders, at every echelon, focus combined arms training on specific integration and synchronization tasks based on their METL. Figure 2-3 illustrates the scope and scale of the combined arms training challenge.

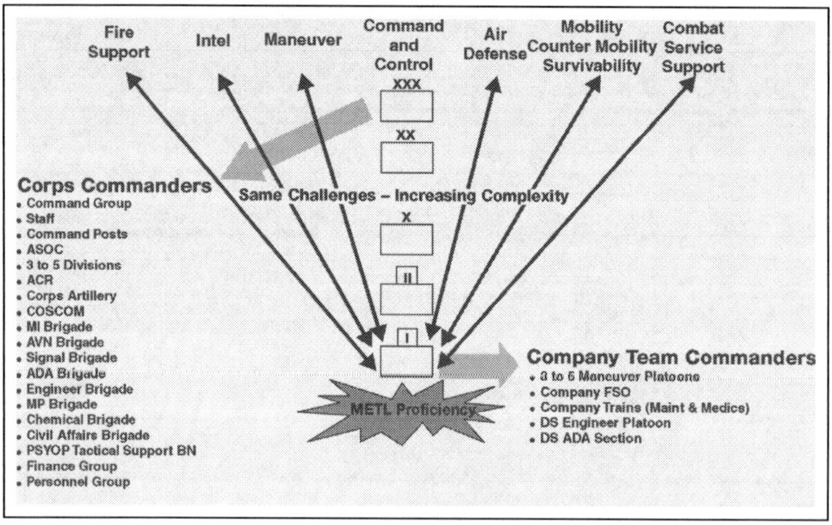

Figure 2-3. Combined Arms Training—Scope and Scale

TRAIN FOR COMBAT PROFICIENCY

2-16. The goal of all training is to achieve the standard. This develops and sustains combat capable warfighting organizations. To achieve this, units must train to standard under realistic conditions. Achieving standards requires hard work by commanders, staff officers, unit leaders, and soldiers. Within the confines of safety and common sense, commanders and leaders must be willing to accept less than perfect results initially and demand realism in training. They must integrate such realistic conditions as imperfect intelligence; reduced communications; smoke; noise; rules of engagement; simulated nuclear, biological, and chemical environments; battlefield debris; loss of key leaders; civilians on the battlefield; JIM requirements; and varying extremes in weather. They must seize every opportunity to move soldiers out of the classroom into the field; fire weapons; maneuver as a combined arms team; and incorporate protective measures against enemy actions. Although CTCs provide the most realistic and challenging training experience in the Army, they must not be viewed as an "end point" in the unit-training life cycle. Rather, they provide a "go to war experience" by which commanders can assess their METL proficiency and determine the effectiveness of their training program.

- **Realistic.** Tough, realistic, and intellectually and physically challenging training excites and motivates soldiers and leaders. Realistic training builds competence and confidence by developing and honing skills, and inspires excellence by fostering initiative, enthusiasm, and eagerness to learn. Successful completion of each training phase increases the capability and motivation of individuals and units for more sophisticated and challenging achievement. This is the commanders' continuous quest.
- **Performance-Oriented.** Units become proficient in the performance of critical tasks and missions by practicing the tasks and missions. Soldiers learn best by doing, using an experiential, hands-on approach. Commanders and subordinate leaders plan training that will provide these opportunities. All training assets and resources, to include training aids, devices, simulators, and simulations (TADSS), must be included in the unit's training strategy.

TRAIN TO STANDARD USING APPROPRIATE DOCTRINE

2-17. Training must be done to the Army standard and conform to Army doctrine. If mission tasks involve emerging doctrine or non-standard tasks, commanders establish the tasks, conditions and standards using mission orders and guidance, lessons learned from similar operations, and their professional judgment. The next higher commander approves the creation of the standards for these tasks. FM 3-0 provides the doctrinal foundations; supporting doctrinal manuals describe common TTP that permit commanders and organizations to adjust rapidly to changing situations. Doctrine provides a basis for a common vocabulary across the force. In units, new soldiers will have little time to learn non-standard procedures. Therefore, units must train to the Army standard contained in the MTP and STPs, while applying Army doctrine and current regulatory guidance. When serving as a joint headquarters and conducting joint training Army organizations use joint doctrine and TTP. Joint doctrine establishes the fundamentals of joint operations and provides guidance on how best to employ joint forces. This linkage between operational and training doctrine is critical to successful training.

TRAIN TO ADAPT

2-18. Commanders train and develop adaptive leaders and units, and prepare their subordinates to operate in positions of increased responsibility. Repetitive, standards-based training provides relevant experience. Commanders intensify training experiences by varying training conditions. Training experiences coupled with timely feedback builds competence. Leaders build unit, staff and soldier confidence when they consistently demonstrate competence. Competence, confidence, and discipline promote initiative and enable leaders to adapt to changing situations and conditions. They improvise with the resources at hand, exploit opportunities and accomplish their assigned mission in the absence of orders. Commanders at every echelon integrate training events in their training plans to develop and train imaginative, adaptive leaders and units.

TRAIN TO MAINTAIN AND SUSTAIN

2-19. Soldier and equipment maintenance is a vital part of every training program. Soldiers and leaders are responsible for maintaining all assigned equipment and supplies in a high state of readiness to support training or operational missions. Units must be capable of fighting for sustained periods of time with the equipment they are issued. Soldiers must become experts in both the operation and maintenance of their equipment. This link between training and sustainment is vital to mission success.

TRAIN USING MULTIECHELON TECHNIQUES

2-20. Multiechelon training is the most effective and efficient way of sustaining proficiency on mission essential tasks with limited time and resources. Commanders use multiechelon training to—

- Train leaders, battle staffs, units, and individuals at each echelon of the organization simultaneously.
- Maximize use of allocated resources and available time.
- Reduce the effects of personnel turbulence.

2-21. Large-scale training events provide an excellent opportunity for valuable individual, leader, crew, and small unit training. Multiechelon training can occur when an entire organization is training on one single METL task or when different echelons of an organization conduct training on related METL tasks simultaneously. (See chapter 4 for detailed discussion on multiechelon training.) All multiechelon training techniques—

- Require detailed planning and coordination by commanders and leaders at each echelon.
- Maintain battle focus by linking individual and collective battle tasks with unit METL tasks, within large-scale training event METL tasks.
- Habitually train at least two echelons simultaneously on selected METL tasks.

TRAIN TO SUSTAIN PROFICIENCY

2-22. Once individuals and units have trained to a required level of proficiency, leaders must structure individual and collective training plans to retrain critical tasks at the minimum frequency necessary to sustain proficiency. Sustainment training is the key to maintaining unit proficiency through personnel turbulence and operational deployments. MTP and individual training plans are tools to help achieve and sustain collective and individual proficiency. Sustainment training must occur often enough to train new soldiers and minimize skill decay. Army units train to accomplish their missions by frequent sustainment training on critical tasks. Infrequent "peaking" of training for an event (CTC rotation, for example) does not sustain wartime proficiency. Battle focused training is training on wartime tasks. Many of the METL tasks that a unit trains on for its wartime mission are the same as required for a stability operation or support operation that they might execute.

2-23. Sustainment training enables units to operate in a Band of Excellence (figure 2-4) through appropriate repetition of critical tasks. The Band of Excellence is the range of proficiency within which a unit is capable of executing its wartime METL tasks. For RC units the Band of Excellence is the range of proficiency within which a unit is capable of executing its premobilization tasks. Training to sustain proficiency in the Band of Excellence includes training leaders, battle staffs, and small lethal units. The solid black line shows the results of an effective unit training strategy that sustains training proficiency over time, maintaining it within the Band of Excellence. The dotted black line shows an ineffective training strategy that often causes the unit to fall outside the Band of Excellence, thus requiring significant additional training before the unit is capable of executing its wartime METL tasks. Personnel turbulence and availability of resources pose a continuous challenge to maintaining METL proficiency within the Band of Excellence.

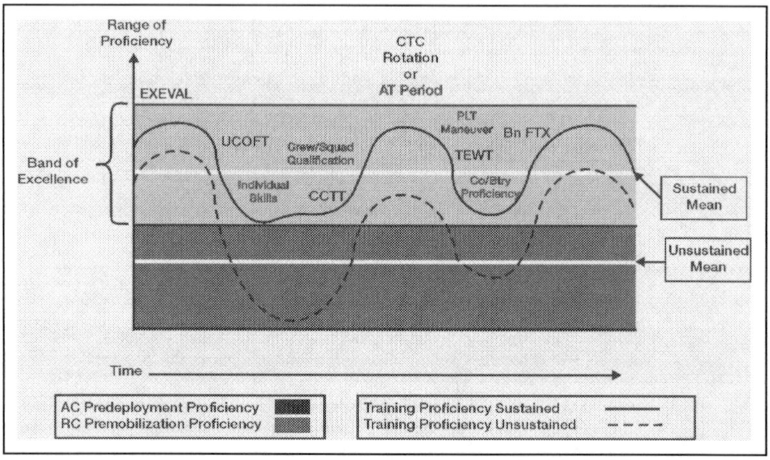

Figure 2-4. Band of Excellence

2-24. The Army provides combat ready forces on short notice to combatant commanders. Units transition from training locations to operational theaters using the train-alert-deploy sequence. Commanders recognize that crises rarely allow sufficient time to correct training deficiencies between alert and deployment. They strive to ensure their units are prepared to accomplish their METL tasks before alert and refine mission specific training in the time available afterwards. Accordingly, applying the principles of training, a commander conducts training to sustain proficiency on METL tasks within the Band of Excellence to ensure mission readiness. Mission specific training can be conducted as organizations are alerted and deployed based on time available.

2-25. RC units require postmobilization training to achieve proficiency at level organized. Postmobilization training time can be minimized by focusing on MOS qualification, and crew, squad, section and platoon proficiency for combat arms, and company, battery, and troop proficiency for CS/CSS units during premobilization training.

TRAIN AND DEVELOP LEADERS

2-26. Commanders have a duty and execute a vital role in leader training and leader development. They teach subordinates how to fight and how to train. They mentor, guide, listen to, and "think with" subordinates. They train leaders to plan training in detail, prepare for training thoroughly, execute training aggressively, and evaluate short-term training proficiency in terms of desired long-term results. Training and developing leaders is an embedded component of every training event. Nothing is more important to the Army than building confident, competent, adaptive leaders for tomorrow.

COMMANDERS AND TRAINING

2-27. Effective training is the number one priority of commanders. The commander is the primary trainer and responsible for the wartime readiness of their formation. In wartime, training continues with a priority second only to combat or to the support of combat operations. Commanders and senior leaders must extract the greatest training value from every training opportunity. Effective training requires the commander's continuous personal time and energy to accomplish the following—

DEVELOP AND COMMUNICATE A CLEAR VISION

2-28. The senior leader's training vision provides the direction, purpose, and motivation necessary to prepare individuals and organizations to win in battle. It is based on a comprehensive understanding of—

- Mission, doctrine, and history.
- Enemy/threat capabilities.
- Operational environment.
- Organizational and personnel strengths and weaknesses.
- Training environment.

TRAIN ONE ECHELON BELOW AND EVALUATE TWO ECHELONS BELOW

2-29. Commanders are responsible for training their own unit and one echelon below. Commanders evaluate units two echelons below. For example, brigade commanders train battalions and evaluate companies; battalion commanders train companies and evaluate platoons.

REQUIRE SUBORDINATES TO UNDERSTAND AND PERFORM THEIR ROLES IN TRAINING

2-30. Since good training results from leader involvement, one of the commander's principal roles in training is to teach subordinate trainers how to train and how to fight. The commander provides the continuing leadership that focuses on the organization's wartime mission. The commander assigns officers the primary responsibility for collective training and NCOs the primary responsibility for individual, crew, and small team training. The commander, as the primary trainer, uses multiechelon techniques to meld leader, battle staff, and individual training requirements into collective training events, while recognizing the overlap in training responsibilities (figure 2-5). Commanders teach, coach, and mentor subordinates throughout.

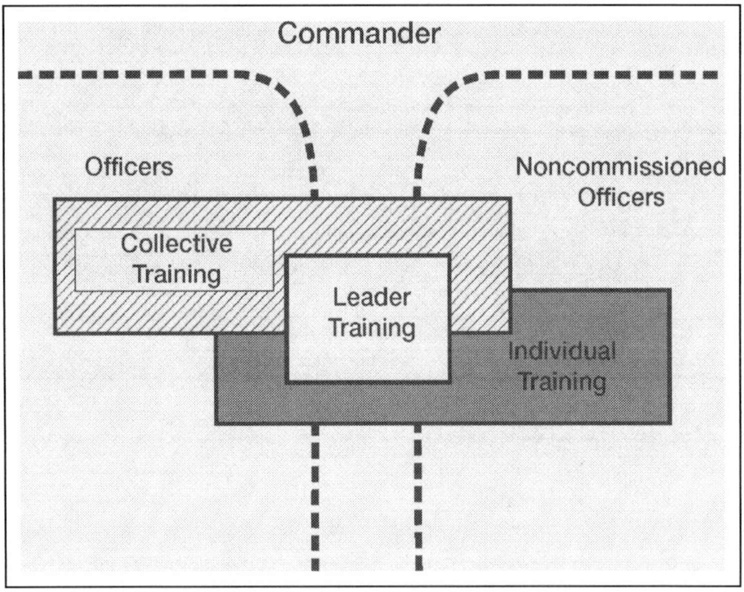

Figure 2-5. Overlapping Training Responsibilities

TRAIN ALL ELEMENTS TO BE PROFICIENT ON THEIR MISSION ESSENTIAL TASKS

2-31. Commanders must integrate and train to Army standard all BOS, within and supporting their command, on their selected mission essential tasks. An important requirement for all leaders is to project training plans far enough into the future and to coordinate resources with sufficient lead time.

DEVELOP SUBORDINATES

2-32. Competent and confident leaders build cohesive organizations with a strong chain of command, high morale, and good discipline. Therefore, commanders create leader development programs that develop warfighter professionalism—skills and knowledge. They develop their subordinates' confidence and empower them to make independent, situational-based decisions on the battlefield. Commanders assist subordinates with a self-development program and share experienced insights that encourage subordinates to study and learn their profession. They train leaders to plan training in detail, prepare for training thoroughly, execute aggressively, and evaluate short-term training proficiency in terms of desired long-term results. Effective leader development programs will continuously influence the Army as junior leaders progress to higher levels of responsibility.

INVOLVE THEMSELVES PERSONALLY IN PLANNING, PREPARING, EXECUTING, AND ASSESSING TRAINING

2-33. The senior commander resources training and protects subordinate commanders' training time. They are actively involved in planning for future training. They create a sense of stability throughout the organization by protecting approved training plans from training distracters. Senior commanders protect the time of subordinate commanders allowing them to be present at training as much as possible. Subordinate commanders are responsible for executing the approved training to standard. Senior commanders are present during the conduct of training as much as possible and provide experienced feedback to all participants.

DEMAND TRAINING STANDARDS ARE ACHIEVED

2-34. Leaders anticipate that some tasks will not be performed to standard. Therefore, they design time into training events to allow additional training on tasks not performed to standard. It is more important to train to standard on a limited number of critical tasks, rather than attempting and failing to achieve the standard on too many tasks, rationalizing that corrective action will occur during some later training period. Soldiers will remember the enforced standard, not the one that was discussed.

ENSURE PROPER TASK AND EVENT DISCIPLINE

2-35. Senior leaders ensure junior leaders plan the correct task-to-time ratio. Too many tasks guarantee nothing will get trained to standard and no time is allocated for retraining. Too many events result in improper preparation and recovery.

FOSTER A COMMAND CLIMATE THAT IS CONDUCIVE TO GOOD TRAINING

2-36. Commanders create a climate that rewards subordinates who are bold and innovative trainers. They challenge the organization and each individual to train to full potential. Patience and coaching are essential ingredients to ultimate achievement of the Army standard.

ELIMINATE TRAINING DISTRACTIONS

2-37. The commander who has planned and resourced a training event is responsible to ensure participation by the maximum number of soldiers. Administrative support burdens cannot be ignored, however, they can be managed using an effective time management system. Senior commanders must support subordinate commanders' efforts to train effectively by eliminating training distracters and reinforcing the requirement for all assigned personnel to be present during training.

TOP-DOWN/BOTTOM-UP APPROACH TO TRAINING

2-38. The top-down/bottom-up approach to training is a team effort in which senior leaders provide training focus, direction and resources, and junior leaders provide feedback on unit training proficiency, identify specific unit training needs, and execute training to standard in accordance with the approved plan. It is a team effort that maintains training focus, establishes training priorities, and enables effective communication between command echelons.

2-39. Guidance, based on wartime mission and priorities, flows from the top-down and results in subordinate units' identification of specific collective and individual tasks that support the higher unit's mission. Input from the bottom up is essential because it identifies training needs to achieve task proficiency on identified collective and individual tasks. Leaders at all echelons communicate with each other about requirements, and planning, preparing, executing, and evaluating training.

2-40. Senior leaders centralize planning to provide a consistent training focus from the top to the bottom of the organization. However, they decentralize execution to ensure that the conduct of mission related training sustains strengths and overcomes the weaknesses unique to each unit. Decentralized execution promotes subordinate leaders' initiative to train their units, but does not mean senior leaders give up their responsibilities to supervise training, develop leaders, and provide feedback.

BATTLE FOCUS

2-41. Battle focus is a concept used to derive peacetime training requirements from assigned and anticipated missions. The priority of training in units is to train to standard on the wartime mission. Battle focus guides the planning, preparation, execution, and assessment of each organization's training program to ensure its members train as they are going to fight. Battle focus is critical throughout the entire training process and is used by commanders to allocate resources for training based on wartime and operational mission requirements. Battle focus enables commanders and staffs at all echelons to structure a training program that copes with non-mission related requirements while focusing on mission essential training activities. It is recognition that a unit cannot attain proficiency to standard on every task whether due to time or other resource constraints. However, commanders can achieve a successful training program by consciously focusing on a reduced number of critical tasks that are essential to mission accomplishment.

2-42. A critical aspect of the battle focus concept is to understand the responsibility for, and the linkage between, the collective mission essential tasks and the individual tasks that support them. The diagram at figure 2-6 depicts the relationships and the proper sequence to derive optimum training benefit from each training opportunity.

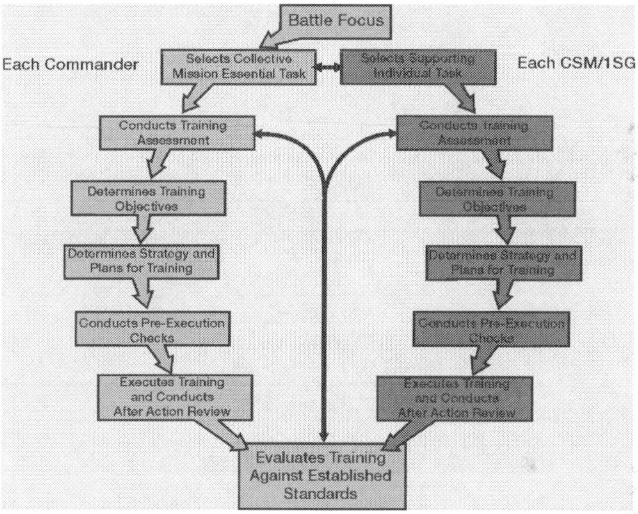

Figure 2-6. Integration of Collective and Individual Training

2-43. The commander and the CSM or 1SG must jointly coordinate the collective mission essential tasks and individual training tasks on which the unit will concentrate its efforts during a given period. The CSM or 1SG must select the specific individual tasks that support each collective task to be trained. Although NCOs have the primary role in training and sustaining individual soldier skills, officers at every echelon remain responsible for training to established standards during both individual and collective training. Battle focus is applied to all missions across the full spectrum of operations.

ARMY TRAINING MANAGEMENT CYCLE

2-44. The foundation of the training process is the Army Training Management Cycle (figure 2-7). In the METL development process (chapter 3), training must be related to the organization's wartime operational plans and focus on METL tasks. The availability of resources does not affect METL development. The METL is an unconstrained statement of the tasks required to accomplish wartime missions. Resources for training, however, are constrained and compete with other missions and requirements. Leaders develop the long-range, short-range, and near-term training plans (chapter 4) to utilize effectively available resources to train for proficiency on METL tasks. After training plans are developed, units execute training by preparing, conducting, and recovering from training (chapter 5). The process continues with training evaluations that provide bottom-up input to organizational assessment. Organizational assessments provide necessary feedback to the senior commander that assist in preparing the training assessment (chapter 6).

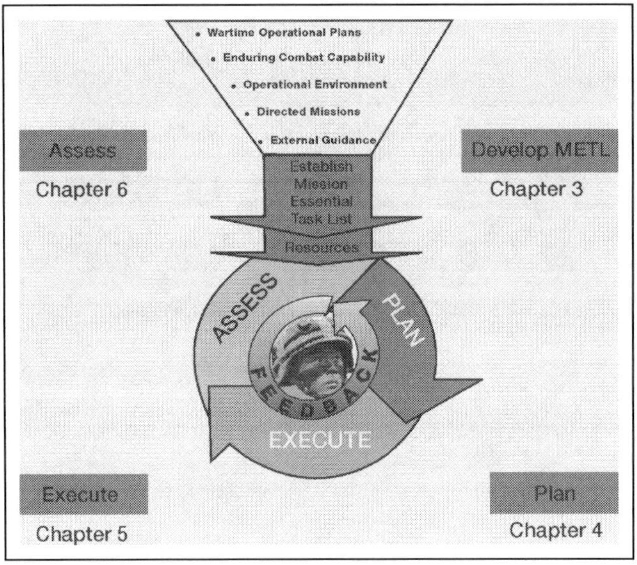

Figure 2-7. Army Training Management Cycle

Chapter 3

Mission Essential Task List Development

Army Mission Essential Tasks

- ❒ *Shape the security environment*
- ❒ *Respond promptly to crisis*
- ❒ *Mobilize the Army*
- ❒ *Conduct forcible entry operations*
- ❒ *Dominate land operations*
- ❒ *Provide support to civil authorities*

FM 1, The Army and FM 3-0, Operations

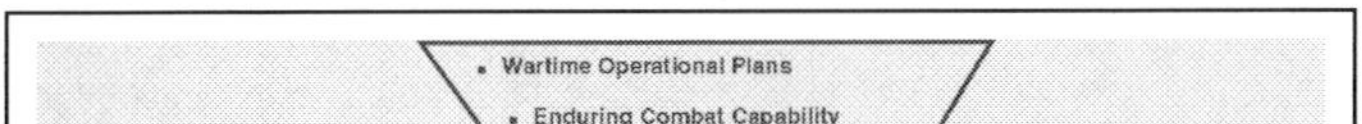

METL

3-1. A mission essential task is a collective task in which an organization has to be proficient to accomplish an appropriate portion of its wartime operational mission. Army organizations, whether they are AC or RC, Modification Table of Organization and Equipment (MTOE) or Table of Distribution and Allowances (TDA), cannot achieve and sustain proficiency on every possible training task. The commander must identify those tasks that are essential to accomplishing the organization's wartime operational mission. Battle-focused METL identifies those tasks that are essential to the accomplishment of the unit's wartime operational mission and provides the foundation for the unit's training program.

3-2. All company level and above units develop a METL that is approved by its designated wartime commander. Detachments, organized with a commander and under a distinct MTOE or TDA, also develop a METL. Explosive ordnance detachments, transportation port operation cargo detachments and preventive medicine medical detachments are examples of these type units.

METL DEVELOPMENT PROCESS

3-3. The METL development process reduces the number of tasks the organization must train and focuses the organization's training efforts on the most important collective training tasks required to accomplish the mission.

3-4. METL development is the catalyst that keeps Army training focused on wartime operational missions. Applying the METL development—

- Focuses the unit's training on essential tasks.
- Provides a forum for professional discussion and leader development among senior, subordinate and adjacent (peer) commanders concerning the linkage between mission and training.
- Enables subordinate commanders and key NCOs to crosswalk collective, leader and individual tasks to the mission.
- Leads to "buy-in" and commitment of unit leaders to the organization's training plan.

3-5. Figure 3-1 depicts the process that commanders use to identify and select mission essential tasks.

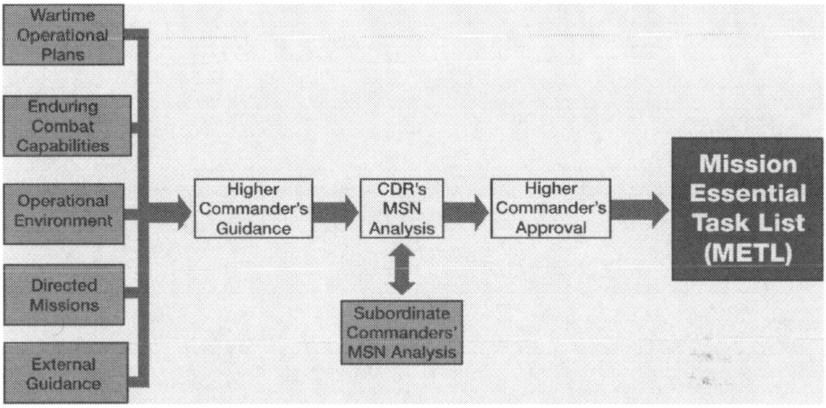

Figure 3-1. METL Development Process

INPUTS TO METL DEVELOPMENT

3-6. There are five primary inputs to METL development.

Wartime Operational Plans

3-7. The most critical inputs to METL development are the organization's wartime operational and contingency plans. The missions and related information provided in these plans are key to determining essential training tasks.

Enduring Combat Capabilities

3-8. The fundamental reason for the organization and operation of Army forces is to generate effects of combined arms in order to contribute to successful execution of wartime operational missions. To do this, Army commanders form combat, CS, and CSS forces into cohesive teams through training for combat proficiency. Enduring combat capabilities are the unique contribution each unit makes to ensure the Army successfully accomplishes any mission anytime anywhere.

Operational Environment

3-9. The operational environment has six dimensions; Threat, Political, Unified Action, Land Combat Operations, Information, and Technology (see FM 3-0). Each dimension affects how Army forces combine, sequence, and conduct military operations. Commanders tailor forces, employ diverse capabilities, and support different missions to succeed in this complex environment.

Directed Missions

3-10. Army organizations are frequently directed to conduct a mission other than its assigned wartime operational mission. These missions range from major combat operations to providing humanitarian assistance or other types of stability and support operations.

External Guidance

3-11. External guidance serves as an additional source of training tasks that relate to an organization's wartime operational mission. Some examples are—

- Higher headquarters directives.
- MTP.
- Force integration plans.
- Army Universal Task List (AUTL).
- Universal Joint Task List (UJTL).

3-12. In some cases, external guidance identifies tasks that make up the mission (for example, MTPs). In others, they specify additional tasks that relate to the mission (for example, mobilization plans, directed stability operations or support operations). Figure 3-2 is an example of brigade tasks derived from the five primary inputs to the unit's METL.

Alert and Deploy the Brigade	
• Draw and Upload Basic/Operational Loads • Conduct Soldier Readiness/Administrative / Logistic Preparation for Overseas Movement • Deploy Advance Parties Or Liaison Officers	• Move by Road or Rail to Aerial Port of Embarkation (APOE) or Seaport of Embarkation (SPOE) • Upload Equipment at APOE or SPOE
Conduct Attack	***Conduct Defense***
• Attack a Moving Enemy • Attack a Stationary Enemy • Movement to Contact	• Conduct a Mobile Defense • Conduct an Area Defense
Conduct Support Operations	***Conduct Stability Operations***
• Domestic Support Operations • Foreign Humanitarian Assistance	• Peacekeeping Operations • Combat Terrorism • Support Counter-Drug Operations
Conduct Sustainment Operations	
• Provide Medical Treatment and Evacuation (air and ground) • Move by Air/Surface Transportation • Manage Terrain	• Recover and Evacuate Disabled Equipment • Control Reconstitution of Subordinate Units • Conduct Mortuary Affairs Operations

Figure 3-2. Example of Brigade Tasks Derived from Wartime Operational Plans and External Guidance

3-13. In similar type organizations, METL may vary significantly because of different missions or geographical locations. For example, a power projection organization may identify strategic deployment requirements as critical deployment tasks while a forward-deployed organization may identify tactical deployment requirements (such as rapid assembly and tactical road marches) as critical deployment tasks. Geography may also influence the selection of different mission essential tasks for units with missions in tropical, cold, or desert environments.

COMMANDER'S ANALYSIS

3-14. The commander's analysis of wartime operational plans, and others primary input to the METL, identify those tasks critical for wartime mission accomplishment. Higher commanders provide guidance to help their subordinate commanders focus this analysis. Commanders coordinate the results of their analysis with subordinate and adjacent commanders. The higher commander approves the METL. This process provides the means to coordinate, link, and integrate a wartime operational mission focused METL throughout the organization.

3-15. To illustrate the METL development process, the following brigade wartime mission statement forms the start point for determining the most important training tasks:

At C-day, H-hour, Brigade deploys: On order, conducts combat operations assigned by higher headquarters.

3-16. The commander reviews the wartime operational mission statement and other primary input to the METL, and identifies all of the training tasks. Together, these five sources provide the total list of possible training tasks. This analysis results in the list at figure 3-2. The commander then narrows down the list of all derived tasks to those tasks critical for mission accomplishment. These tasks become the brigade's METL. Figure 3-3 shows an example of a brigade METL.

- ❒ Alert and Deploy the Brigade
- ❒ Conduct Attack
  - ▪ Movement to Contact
- ❒ Conduct Area Defense
- ❒ Conduct Sustainment Operations
- ❒ Command and Control Tactical Operations

Figure 3-3. Example of Brigade METL

RESERVE COMPONENT METL DEVELOPMENT

3-17. The METL development process is the same for AC and RC organizations. RC METL development recognizes that RC units have less than 20 percent of the training time available to their AC counterparts. Therefore, battle focus is essential so that RC commanders can concentrate their time on the most critical wartime training requirements. RC units often operate under a chain of command different from their wartime chain of command. The associate AC chain of command assigns missions, provides wartime mission guidance, and approves METLs. The state adjutant general or regional support groups review and coordinate RC METLs. They resource training and ensure that mission training tasks are executed and evaluated. Continental U.S. Armies (CONUSAs) approve the METL for selected RC units (ARNG divisions, enhanced separate brigades, roundout units, reinforcing aviation units, and force support package units with latest arrival dates less than D+30). The peacetime chain of command approves the remainder of RC unit METLs.

ECHELON ABOVE DIVISION/ECHELON ABOVE CORPS (EAD/EAC) METL DEVELOPMENT

3-18. In a similar manner, commanders of EAD/EAC organizations must use the battle focus concept and METL development process to focus their training. Figure 3-4 shows an example of a corps support battalion METL.

- ❒ Deploy/Relocate the Battalion
- ❒ Plan CSS Operations
- ❒ Establish and Sustain Area of Operations
- ❒ Provide CSS
- ❒ Protect the Force

Figure 3-4. Example of Corps Support Battalion (EAD) METL

TDA METL DEVELOPMENT

3-19. Battle focus is equally applicable to TDA organizations. Senior leaders of TDA organizations derive METL from critical peacetime or wartime missions. Mission essential tasks may be either critical training tasks or operational activities. In short, they represent the tasks required to accomplish the TDA organization's mission. Figure 3-5 shows an example of a garrison support unit METL.

- ❒ Alert, Assemble and Mobilize
- ❒ Command and Control General Support Unit, Mobilizing and Demobilizing Units
- ❒ Augment BASOPS Support
- ❒ Provide Mobilization and Demobilization Processing Support
- ❒ Conduct Force Protection Operations

Figure 3-5. Example of Garrison Support Unit (TDA) METL

METL DEVELOPMENT FOR DIRECTED MISSIONS

3-20. When an organization is directed to conduct a mission other than its assigned wartime operational mission (such as a stability operation or support operation), the training management cycle still applies. Directed missions can span the full spectrum of operations. For MTOE organizations, directed missions could range from major combat operations to providing humanitarian assistance or other types of stability operations and support operations. For TDA organizations, directed missions can range from mobilization to installation force protection operations.

3-21. Using their wartime METL as the foundation, commanders who are directed to change their mission conduct a mission analysis, identify METL tasks, and assess training proficiency for the directed mission. The mission analysis of the newly assigned mission could change the unit's METL, training focus, and the strategy to achieve proficiency for METL tasks. Figure 3-6 shows an example of tasks supporting a directed mission involving a stability operation.

Collective Training	
• Convoy Operations • Route Security • Rail/Air Movement Training	• Area Security • Patrolling Operations • Establish/Operate Checkpoints
Leader Training	
• Fire Control Exercise (FCX) • Casualty Evacuation (CAS EVAC) • Deployment Exercise • Risk Management	• Rules Of Engagement (ROE) Proficiency • Petroleum, Oils and Lubricants/Military (POL/MIL) Seminar • Media Interaction
Individual Training	
• Mine Awareness • ROE Proficiency • Media Interaction	• Medical Awareness • Country Orientation • Force Protection

Figure 3-6. Examples of METL Tasks to Support a Directed Mission to Conduct Stability Operation

3-22. In cases where mission tasks involve emerging doctrine or non-standard tasks, commanders establish tasks, conditions, and standards using mission orders and guidance, lessons learned from similar operations, and their professional judgment. Senior commanders approve the established standards for these tasks as part of the normal METL approval process. If time permits prior to deployment, units should execute a mission rehearsal exercise (MRE) with all participating units.

3-23. Upon redeployment from a directed mission, commanders conduct a mission analysis consistent with the training management cycle to reestablish proficiency in the unit's wartime operational METL. Senior commanders must take into account the additional time this reintegration process may take. Battle focus guides the planning, preparation, execution, and assessment of each organization's training program to ensure its members train as they will fight.

JOINT METL (JMETL) DEVELOPMENT

3-24. Army organizations often provide Army forces within joint force formations. The missions and JMETL of such formations are derived from the Universal Joint Task List by the joint force commander and service component commanders, and are approved by the combatant commander.

3-25. A selected Army headquarters may be designated as a JTF headquarters, joint forces land component headquarters (JFLC) or Army Forces (ARFOR) headquarters. This requires the designated Army headquarters to develop a JMETL. The Army headquarters commander crosswalks the JMETL with the current Army headquarters and subordinate unit METLs. Joint training manuals provide an overview of the joint training system (JTS), and assists in—

- Developing joint training requirements.
- Planning joint training.
- Executing joint training.
- Assessing joint proficiency.

METL DEVELOPMENT FUNDAMENTALS

3-26. The following fundamentals apply to METL development—

- The METL is derived from the organization's wartime plans and related tasks in external guidance.
- Mission essential tasks must apply to the entire organization. METL does not include tasks assigned solely to subordinate organizations.
- Each organization's METL must support and complement the METL of higher headquarters.
- The availability of resources does not affect METL development. The METL is an unconstrained statement of the tasks required to accomplish wartime missions.
- Commanders direct operations and integrate the BOS through plans and orders. The BOS are used to systematically ensure that interdependent organizational tasks necessary to generate, sustain, and apply combat power are directed toward accomplishing the overall mission. The BOS are the physical means (soldiers, organizations, and equipment) used to accomplish the mission. The BOS are—
  - **Intelligence.** The intelligence system plans, directs, collects, processes, produces, and disseminates intelligence on the threat and the environment; performs intelligence preparation of the battlefield; and other intelligence tasks. Intelligence is developed as part of a continuous process and is fundamental to Army operations.
  - **Maneuver.** Commanders maneuver forces to create the conditions for tactical and operational success. Maneuver involves movement to achieve positions of advantage with respect to enemy forces. Through maneuver, friendly forces gain the ability to destroy enemy forces or hinder enemy movement by direct and indirect application of firepower or threat of its application.

- **Fire Support.** Fire support consists of fires that directly support land, maritime, amphibious, and special operations forces in engaging enemy forces, combat formations, and facilities in pursuit of tactical and operational objectives. Fire support integrates and synchronizes fires and effects to delay, disrupt, or destroy enemy forces, systems, and facilities. The fire support system includes the collective and coordinated use of target acquisition data, indirect fire weapons, fixed-winged aircraft, electronic warfare, and other lethal and nonlethal means to attack targets.
- **Air Defense.** Air defense protects the force from air and missile attack and aerial surveillance. It prevents enemies from interdicting friendly forces while freeing commanders to synchronize maneuver and fire power. The weapons of mass destruction and proliferation of missile technology increase the importance of the air defense systems.
- **Mobility/Counter-mobility/Survivability.** Mobility operations preserve the freedom of maneuver for friendly forces. Mobility missions include breaching obstacles, increasing battlefield circulation, improving or building roads, providing bridge and raft support, and identifying routes around contaminated areas. Counter-mobility denies mobility to enemy forces. Survivability operations protect friendly forces from the effects of enemy weapon systems and from natural occurrences. Nuclear, biological, and chemical defense measures are essential survivability tasks.
- **Combat Service Support.** Combat service support (CSS) provides the physical means with which forces operate, from the production base and replacement centers in the continental United States to soldiers engaged in close combat. CSS includes many technical specialties and functional activities. It includes maximizing the use of host nation infrastructure and contracted support.
- **Command and Control.** Command and control (C2) has two components—the commander and the C2 system. The C2 system supports the commander's ability to make informed decisions, delegate authority, and synchronize the BOS. Moreover, the C2 system supports the commander's ability to adjust plans for future operations, even while focusing on current operations. Reliable communications are central to C2 systems. Staffs work within the commander's intent to direct units and control resource allocations. Through C2, commanders initiate and integrate all BOS toward a common goal—mission accomplishment.

3-27. Staff elements at each headquarters develop a METL to address mission essential tasks in their areas of responsibility. Figure 3-7 shows a sample Brigade Staff METL. In addition to staff METLs, organizations may develop a METL for each separate command post (for example tactical, main, and rear). The organization's commander or chief of staff approves the staff METL.

Element/Section	Staff Mission Essential Task
S1	Provide Personnel Services Support
S2	Conduct Intelligence Preparation of the Battlefield Develop a Reconnaissance and Surveillance (R&S) Plan
S3	Plan Operations using Military Decision Making Process (MDMP) Prepare OPLAN/OPORD Develop Task Organization Annex and Operations Overlay
S4	Develop Logistics Estimate
S5	Coordinate Host Nation Support
S6	Establish Information Network
DECOORD	Coordinate Fire Support
ENGR	Conduct M/CM/S Operations
ADA LNO	Conduct ADA Planning and Coordination
AVN LNO	Conduct Army Aviation Planning and Coordination
USAF ALO	Conduct A2C2 Management and Control Activities

Figure 3-7. Example of Brigade Staff METL

3-28. Organizations that conduct daily support functions also prepare a METL. The METL for these support organizations must address the differences between peacetime and wartime operating conditions. For example, a CSS unit may operate during peacetime from a permanent facility with some major supplies provided via contract transportation and automation systems operated using commercial telephone systems. A wartime environment, however, may require support missions to be accomplished under austere conditions on an active battlefield.

3-29. The METL for units habitually task organized must be coordinated during the development process. This requirement reinforces the training fundamental that combined arms teams will train as they fight. A key component of the senior commander's METL approval process is determining if each subordinate organization has properly coordinated its METL. A support organization's METL must identify these wartime requirements and include them in their training plans.

METL LINKED TRAINING STRATEGY

3-30. The METL provides the foundation for the organization's training plans. The METL is stabilized once approved. The commander is responsible for developing a training strategy that will maintain unit proficiency for all tasks designated as mission essential.

3-31. Commanders involve subordinate commanders and their CSM/1SG in METL development to create a team approach to battle focused training. Subordinate participation develops a common understanding of the organization's critical wartime requirements so METLs throughout the organization are mutually supporting. Subordinate commanders can subsequently apply insights gained during preparation of the next higher headquarters' METL to the development of their own METL. The CSM/1SG and key NCOs must understand the organization's collective METL so that they can integrate individual tasks into each collective mission essential task during METL based training.

3-32. After the commander designates the collective mission essential tasks required to accomplish the organization's wartime operational mission, the CSM/1SG, in conjunction with key NCOs, develop a supporting individual task list for each mission essential task. Soldier training publications and MTPs are major source documents for selecting appropriate individual tasks.

3-33. There should be no attempt to prioritize tasks within the METL. All METL tasks are equally essential to ensure mission accomplishment. However, all tasks may not require equal training time or resources. The commander allocates training resources to ensure the organization's METL proficiency remains within the Band of Excellence.

3-34. Commanders realize when allocating training time and resources that there are some non-mission related requirements that are critical to the health, welfare, individual readiness, and cohesiveness of a well trained unit. Commanders must carefully select, in conjunction with the CSM/1SG, which non-mission related requirements are critical to the unit. They emphasize the priority of METL training and find opportunities to include non-mission related requirements in the training plan.

3-35. Commanders develop effective training strategies when they crosswalk collective, leader and individual tasks to each METL task with subordinate commanders, CSMs/1SGs, and other key officer and NCO leaders.

TRAINING OBJECTIVES

3-36. After mission essential tasks are selected, commanders identify supporting training objectives for each task. The resulting training objective consists of—

- Task. A clearly defined and measurable activity accomplished by organizations and individuals.
- Condition(s). The circumstances and environment in which a task is to be performed.
- Standard. The minimum acceptable proficiency required in the performance of a particular training task.

3-37. The conditions and standards for many major collective training tasks are identified in applicable MTPs. Figure 3-8 shows an example of a brigade training objective.

MISSION ESSENTIAL TASK:	Conduct an attack.
CONDITIONS:	The brigade is conducting operations independently or as part of a division or Army forces (ARFOR) and has received an operation order (OPORD) or fragmentary order (FRAGO) to conduct an attack at the location and time specified. Coalition forces and noncombatants may be present in the operational environment.
STANDARDS:	1. Brigade leaders gain and or maintain situational awareness (SA). Brigade commander and staff receive an order or anticipate a new mission and begin the military decision-making process (MDMP). Brigade task organizes forces within the brigade. 2. Effects coordination cell (ECC) obtains guidance from the commander; plans, coordinates, and achieves the desired effects utilizing organic and attached assets. 3. Staff plans mobility, countermobility, and survivability: nuclear, biological, and chemical support; air defense (AD) support; and CSS supporting operations. 4. Brigade commander and staff conduct risk management process. 5. Brigade commander and staff conduct backbriefs and rehearsals to ensure subordinates understand commander's intent and concept. 6. Brigade executes the attack; masses all available combat power to destroy enemy per the commander's intent. 7. Brigade consolidates and reorganizes, as necessary. 8. Brigade continues operations, as necessary.

Figure 3-8. Example of Training Objective for a Brigade Mission Essential Task

3-38. The following are documents that will assist commanders and staffs in developing collective and individual training objectives—

- MTP.
- Soldiers manuals.
- Soldier training publications.
- DA Pam 350-38.
- Deployment or mobilization plans.
- AUTL.
- UJTL.
- Army, MACOM, and local regulations.
- Local standing operating procedures (SOP).

BATTLE TASKS

3-39. After review and approval of subordinate organizations' METL, the senior commander selects battle tasks. A battle task is a staff or subordinate organization mission essential task that is so critical that its accomplishment will determine the success of the next higher organization's mission essential task. Battle tasks are selected for each METL task. Battle tasks allow the senior commander to define the training tasks that—

- Integrate the BOS.
- Receive the highest priority for resources such as ammunition, training areas and facilities (to include live and virtual simulators and constructive simulations), materiel, and funds.
- Receive emphasis during evaluations directed by senior headquarters.

3-40. Figure 3-9 shows an example of a division's major subordinate command and separate battalion battle tasks that support the division METL.

Corps Mission Essential Task: Conduct Shaping Operations *Mission Essential Task: Conduct Deliberate Attack*		
Battle Task	***Unit Task Selected***	***Battlefield Operating Systems***
Conduct Deliberate Attack	1st and 2d Brigade	Maneuver
Conduct Deep Attack	Aviation Brigade	Maneuver
Provide Fire Support for Combat Operations	DIVARTY	Fire Support
Reduce Complex Obstacles for Division Deliberate Attack	Engineer Brigade	Mobility/Counter-mobility/Survivability
Provide and Manage DS CSS	DISCOM	CSS
Conduct Intelligence and Electronic Attack Operations	MI Battalion	Intelligence
IOM Tactical Communications Networks (C4I) for the Division	Signal Battalion	Command and Control
Coordinate Air Defense C3I and Directed Early Warning	ADA Battalion	Air Defense

Figure 3-9. Example List of Division MSC and Separate Battalion Battle Tasks That Support a Mission Essential Task

3-41. Figure 3-10 depicts the relationship between wartime missions, METL, and battle tasks. This diagram illustrates how battle focus provides a common direction for the entire organization and the foundation for the subsequent development of relevant training plans.

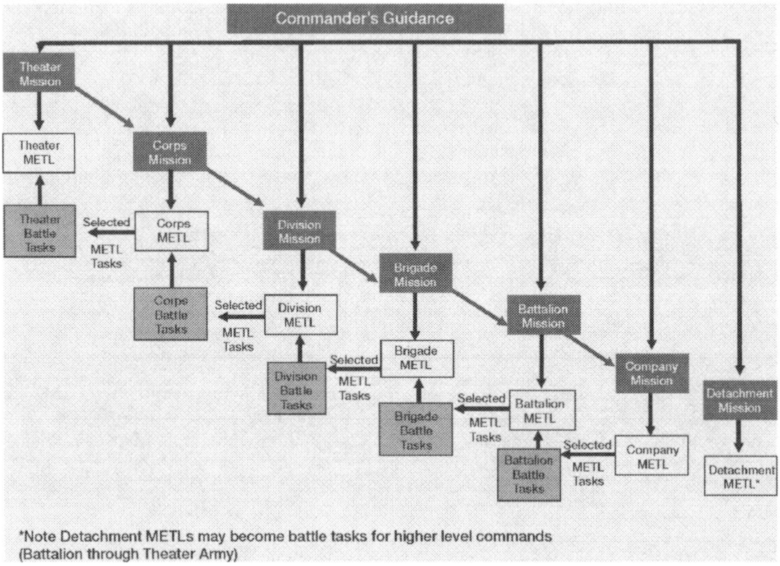

Figure 3-10. Relationships between Mission, METL, and Battle Task

Chapter 4

Planning

We cannot train without planning and we cannot teach without preparation.

General George C. Marshall

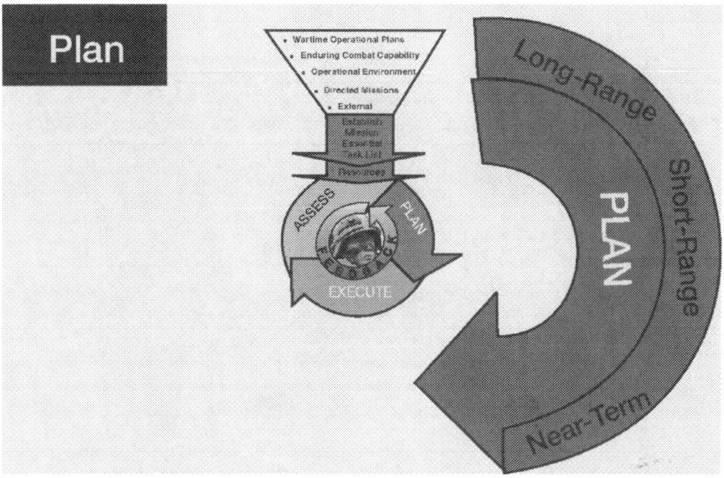

PLANNING PROCESS

4-1. Planning is an extension of the battle focus concept that links organizational METL with the subsequent preparation, execution, and evaluation of training. A relatively centralized process, planning develops mutually supporting METL based training at all echelons within an organization. The planning process ensures continuous coordination from long-range planning, through short-range and near-term planning, and ultimately leads to training execution. The commander's assessment provides direction and focus to the planning process. (Commander's assessment is discussed in chapter 6.) Figure 4-1 depicts the training planning process used to develop battle focused training programs.

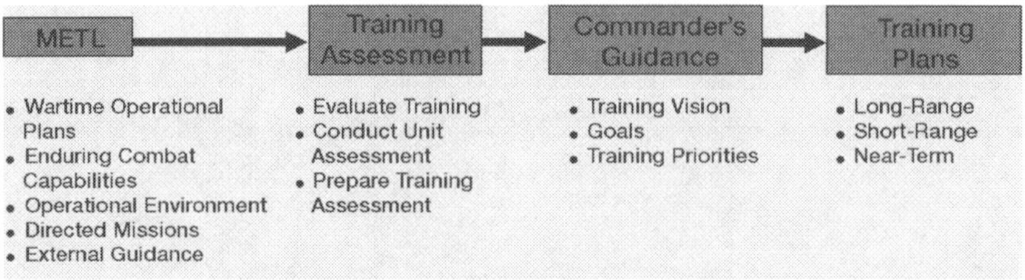

Figure 4-1. Training Planning Process

4-2. The commander applies two principal inputs at the start of the planning process—the METL and the training assessment. Commanders identify tasks that support the METL. The training assessment compares the organization's current level of training proficiency with the desired level of warfighting proficiency. This desired level is defined in MTPs and other doctrinal literature. Commanders may make conscious decisions, based on their training assessment, to defer training for some tasks in which they are currently well trained.

4-3. Leaders determine current training proficiency levels by analyzing all available training evaluations. However, each evaluation applies only to a portion of the total proficiency of an organization at a specific time. Therefore, commanders must use all available evaluation data to develop an assessment of the organization's overall capability to accomplish each mission essential task. In addition to past training evaluations, other information about future events influences the assessment. For example, the projected personnel turnover rates or the fielding of new equipment could significantly affect the commander's assessment of training proficiency status during the upcoming training period. Commanders update the training assessment at the beginning of each long-range and short-range planning cycle and after a major training event or deployment.

4-4. The commander uses the broad experience and knowledge of key subordinates to help determine the organization's current proficiency. Although subordinates provide their evaluation as input for consideration, only the commander can assess the unit's training proficiency. For example, a division commander may direct that the assistant division commanders, key staff members, and subordinate commanders evaluate the training proficiency of the division's ability to execute mission essential tasks and supporting battle tasks. The division CSM and subordinate CSMs evaluate proficiency on individual tasks that support collective tasks. The participants review available collective and individual evaluation information, relying heavily on personal observations. They then compare the organization's current task proficiency with the Army standard. The commander uses subordinate input in making the final determination of the organization's current proficiency on each task (figure 4-2). Commanders assess current METL task proficiency by rating each task as—

- "T" (trained)—The unit is trained and has demonstrated its proficiency in accomplishing the task to wartime standards.
- "P" (needs practice)—The unit needs to practice the task. Performance has demonstrated that the unit does not achieve the standard without some difficulty or has failed to perform some task steps to standard.
- "U" (untrained)—The unit cannot demonstrate an ability to achieve wartime proficiency.

Mission Essential Task	Current Training Status							Overall Task Assessment	Strategy to Improve or Sustain Training Proficiency to Desired Warfighting Levels
	INT	MAN	FS	ADA	M/CM/S	CSS	C2		
Alert and Deploy	P	P	P	P	P	T	P	P	• Exercise unit emergency deployment and readiness procedures during division quarterly EDRE. • Exercise unit alert recall procedures monthly. • Continue semi-annual leader's reconnaissance of the installation's rail, convoy, air and sea deployment facilities. • Continue semi-annual TEWT of installation ammunition supply point (ASP) upload procedures.
Conduct Deliberate Attack	P	T	P	P	P	P	P	P	• Train quarterly during Division CPX or MAPEX • Request corps response cell for each Division exercise to improve C2 • Train annually during division CPX • More emphasis on intelligence and CSS

LEGEND
T Trained
P Needs Practice
U Untrained

Figure 4-2. Extract from Commander's Training Assessment

4-5. The training requirement is the training necessary to achieve and sustain METL task proficiency within the Band of Excellence.

4-6. The commander, assisted by staff, develops a strategy to accomplish each training requirement. This includes improving proficiency on some tasks and sustaining performance on others. Through the training strategy, the commander establishes training priorities by determining the minimum frequency each mission essential task will be performed during the upcoming planning period. The strategy also includes broad guidance that links the METL with upcoming major training events. The initial training assessment includes the commander's guidance that starts the detailed planning process.

4-7. The training assessment of each separate mission essential task enables the commander to develop the commander's training vision. This is a broad concept for training the organization to achieve and sustain wartime proficiency. The key elements that shape a commander's training vision are a thorough understanding of training and operations doctrine, assessment of METL proficiency levels, and knowledge of potential enemy capabilities. The commander's training vision is supported by organizational goals that provide a common direction for all the commander's programs and systems.

4-8. Senior commanders involve their staffs and subordinate commanders in goal development to ensure common understanding and create an organizational team approach. Examples of organizational goals include—

- Establish and support a command climate conducive to developing a high level of individual, leader, and collective warfighting proficiency.
- Conduct force integration while continuously maintaining the short-term readiness of the organization.
- Develop and integrate the standard operating procedures required to employ combined arms teams that can fight and win on the battlefield.
- Recruit and retain high-quality soldiers and leaders.

4-9. Through the training planning process, the commander's guidance (training vision, goals, and priorities) is melded together with the METL and the training assessment into manageable training plans.

TRAINING PLANS

4-10. Figure 4-3 compares the three types of training plans—

- Long-range.
- Short-range.
- Near-term.

Long-Range	Short-Range	Near-Term
Disseminate METL and battle tasks. Establish training objective for each mission essential task. Schedule projected major training events. Identify long lead time resources and allocate major resources such as major training area rotations. Identify available training support system products and services and identify new requirements. Coordinate long-range calendars with all supporting agencies to eliminate training detractors. Publish long-range guidance and planning calendar. Provide basis for command operating budget input. Provide long-range training input to higher headquarters.	Refine and expand upon appropriate portions of long-range plan. Cross reference each training event with specific training objectives. Identify and allocate short lead time resources such as local training facilities. Coordinate short-range calendar with all support agencies. Publish short-range guidance and planning calendar. Provide input to unit training meetings.	Refine and expand upon short-range plan through conduct of training meetings. Determine best sequence for training. Provide specific guidance for trainers. Allocate training support system products and services, including training aids, devices, simulators, simulations, and similar resources to specific trainers. Publish detailed training schedules. Provide basis for executing and evaluating training.

Figure 4-3. Comparison of Long-Range, Short-Range, and Near-Term Training Plans

4-11. Properly developed training plans will—

- **Maintain a consistent battle focus.** Each headquarters in the organization involves its subordinate headquarters in the development of training plans. Based on the higher headquarters' plans, subordinate commanders prepare plans that have a battle focus that is consistent throughout the organization.
- **Be coordinated with habitually task organized supporting organizations.** Brigade combat team and battalion task force commanders plan for coordinated combined arms training of their wartime task organizations. Commanders of habitually task-organized units actively participate in this process and develop complementary training plans. Corps and division commanders require integrated training plans and monitor coordination efforts during the planning process.
- **Focus on the correct time horizon.** Long-range training plans in the AC extend out at least one year. The RC long-range plans consider a minimum of two years. Short-range training plans in the AC normally focus on an upcoming quarter (three months) while RC short-range training plans typically use a one-year planning horizon. Near-term planning

for the AC starts approximately eight weeks prior to the execution of training with the RC starting approximately four months prior.

- **Be concerned with future proficiency.** Training plans must focus on raising or sustaining the proficiency of mission essential tasks to the Army standard.
- **Incorporate risk management into all training plans.** The nature of the military profession is inherently dangerous. Commanders must train their units to tough standards under the most realistic conditions possible. Application of the risk management process will not detract from this training goal, but will enhance execution of highly effective, realistic training. Risk management is the process of identifying, assessing, and controlling risks arising from operational factors and making decisions that balance risk costs with mission training benefits. Leaders and soldiers at all echelons use risk management to conserve combat power and resources. Leaders and staffs continuously identify hazards and assess both accident and tactical risks. They then develop and coordinate control measures to mitigate or eliminate hazards. Risk management is a continuous process for each mission or training event. It must be integral to military decisions, tied into each training plan, and become a continuous part of preparation for training.
- **Establish organizational stability.** Changes disrupt training and frustrate subordinate leaders and soldiers. Planning allows organizations to anticipate and incorporate change in a coordinated manner. Stability and predictability are the result of locking in training plans. Senior commanders are responsible to protect subordinate units from change.
- **Make the most efficient use of resources.** The planning process allocates limited time and other resources for training that contributes most to achieving and sustaining wartime proficiency levels.

LONG-RANGE PLANNING

4-12. Senior commanders publish their training guidance document sufficiently in advance to provide adequate planning time for both their wartime units and supporting peacetime organizations. Guidance at these senior command echelons is critical to the development and integration of a large number of subordinate AC and RC long-range training plans. Therefore, long lead times are the norm. The long-range planning cycles for MACOM, corps, AC and RC divisions and subordinate headquarters are at figures 4-4 and 4-5. Each headquarters follows these time lines to allow subordinates adequate time to prepare their plans.

Action[1]	*Planning Guidance Publication Date*[2]	*Future Planning Horizon*
MACOM publishes training guidance and major event calendar	18 months prior to start of a 2-year period[3]	Up to 10 years or more
Corps publishes training guidance and major event calendar	12 months prior to start of a 2-year period[3]	5 to 7 years
Division, separate brigade, regiment, and separate group publish command training guidance and long-range calendar[4]	8 months prior to FY start	Command training guidance at least 1 year Calendar at least 2 years
Installation and community publish long-range calendar	7 months prior to FY start	At least 1 year
Brigade and group publish command training guidance and long-range calendar	6 months prior to FY start	Command training guidance at least 1 year Calendar at least 18 months
Battalion, squadron, and separate company publish long-range calendar	4 months prior to FY start	At least 1 year

[1]These actions also apply to similar command level TDA organizations or activities. For example, a TRADOC school normally commanded by a major general follows the same planning cycle as a division commander.

[2]Each headquarters follows this time line to allow subordinates adequate time to prepare their plans.

[3]Updated annually at the discretion of the commander.

[4]Division, separate brigade, regiment, and separate group commanders normally brief to and receive approval of the next higher headquarters on their long-range training plans no later than 8 months prior to FY start.

Figure 4-4. Active Component Long-Range Planning Cycle

Action[1 2]	Planning Guidance Publication Date[3]	Future Planning Horizon
Division, separate brigade, regiment, and separate group publish command training guidance and long-range calendar[4]	12 months prior to FY start	Command training guidance at least 2 years Calendar at least 5 years
Brigade and separate battalion publish command training guidance and long-range calendar	10 months prior to FY start	At least 5 years
Battalion, squadron, and separate company publish long-range calendar	6 months prior to FY start	At least 3 years

[1]These actions also apply to similar command level TDA organizations or activities. For example, a regional support command commanded by a major general follows the same planning cycle as a division commander.

[2]Intermediate headquarters, such as Continental U.S. Armies, State Area Commands, Major U.S. Army Reserve Commands, and General Officer Commands, provide training guidance and scheduling information in sufficient time to allow subordinate units to meet required publication dates.

[3]Each headquarters follows this time line to allow subordinates adequate time to prepare their plans.

[4]Division, separate brigade, regiment, and separate group commanders normally brief to, and receive approval of, the next higher headquarters in the peacetime chain of command 8 to10 months prior to FY start.

Figure 4-5. Reserve Component Long-Range Planning Cycle

COMMAND TRAINING GUIDANCE (CTG)

4-13. The CTG is published to document the organization's long-range training plan. It is the training analog of the organization's war plan. It must be read and understood by all commanders, staff officers, and senior NCOs. The CTG is used as a ready reference for the planning, preparation, execution, and evaluation of training throughout the long-range planning period. Examples of topics normally addressed in the CTG are—

- Commander's training philosophy.
- METL and associated battle tasks.
- Combined arms training.
- JIM training, as applicable.
- Major training events and exercises.
- Organizational Inspection Program (OIP).
- Leader training.
- Battle staff training.
- Individual training.
- Self development.
- Standardization.
- Training evaluation and feedback.
- New equipment training and other force integration considerations.
- Resource allocations.
- Training management.
- Risk management.

LONG-RANGE PLANNING CALENDAR

4-14. Commanders publish the long-range planning calendar concurrently with their CTG. The calendar graphically depicts the schedule of events described in the CTG. Any known major training events or deployments scheduled beyond the normal planning window appear on the long-range planning calendar. To provide extended planning guidance for RC organizations, AC and RC planners routinely forecast major events that require RC participation for up to five years into the future. They include major events, such as annual training periods and overseas deployments for training (ODT), on their long-range calendars. Upon publication and approval by higher headquarters, long-range planning calendars are "locked in" to provide planning stability to subordinate organizations. This means that only the approving commander can change a long-range planning calendar. The senior commander agrees to allocate and protect the requisite resources, including time, and the subordinate commanders agree to conduct training to standard in accordance with the published calendar.

4-15. Commanders coordinate long-range planning calendars with subordinate commanders, installation support agencies, and any other organizations that can generate training distracters if not fully integrated into the training organization's long-range plan.

4-16. Senior leaders at all echelons eliminate nonessential activities that detract from METL based training. In peacetime, however, certain activities occur that do not directly relate to an organization's wartime mission but are important to other Army priorities. An example of this is AC support of ROTC summer training; for the RC, state-directed requirements for Army National Guard units. Senior leaders limit these peacetime activities to the maximum extent possible. Those that are absolutely essential are included in long-range planning documents. When assigned these activities, commanders continually seek mission related training opportunities.

4-17. During long-range planning, commanders organize training time to support METL training and concentrate training distracters in support periods. In addition to individual requirements such as leave and medical appointments, units may have temporary duty details and other support functions at the installation level. Failure to consider these requirements early in the planning process can cause disruption to planned mission essential training.

TRAINING AND TIME MANAGEMENT

4-18. The purpose of time management is to achieve and sustain technical and tactical competence and maintain training proficiency within the Band of Excellence. Time management systems identify, focus, and protect prime time training periods and the resources to support the training so subordinate organizations are able to concentrate on mission essential training. Figure 4-6 describes a Green-Amber-Red time management system and lists some of the training and support concepts that generally characterize each period. Specific activities will vary between installations according to the local situation and requirements. Time management periods are depicted on applicable long-range planning calendars.

- *Green.* The training focus of organizations in Green periods is multiechelon, collective training that leads to METL proficiency. This period coincides with the availability of major training resources and key training facilities and devices. Organizations in Green periods conduct planned training without distraction and external taskings.
- *Amber.* The focus of units in Amber periods is on training proficiency at the platoon, squad, and crew level. Individual self-development is maximized through the use of installation education centers and through distributed learning. Organizations in Amber periods are assigned support taskings beyond the capability of those units in the Red period, but commanders strive for minimal disruption to Amber organizations' training programs.
- *Red.* The training focus of units in Red periods is on maximizing self-development opportunities to improve leader and individual task proficiency. Units in Red periods execute details and other administrative requirements and allow the maximum number of soldiers to take leave. Block leave is a technique that permits an entire unit to take leave for a designated period of time. Commanders maintain unit integrity when executing administrative and support requirements. This exercises the chain of command and provides individual training opportunities for first-line leaders.

➢*Green Cycle*

- Training focus primarily on collective tasks with individual and leader tasks integrated during multiechelon training.
- Maximum soldier attendance at prime time, mission essential training.
- Coincides with availability of major resources and key training facilities or devices.
- Administrative and support requirements that keep personnel from participating in training eliminated to the maximum extent possible.
- Leaves and passes limited to the minimum essential.

➢*Amber Cycle*

- Small unit, crew, leader and individual soldier training emphasized.
- Provides time for soldier attendance at education and training courses.
- Some sub-organizations may be able to schedule collective training.
- Scheduling of periodic maintenance services.
- Selected personnel diverted to support requirements when all available personnel in organizations in red period are completely committed to support requirements.

➢*Red Cycle*

- Maximize self development.
- Diverts the minimum essential number of personnel to perform administrative and support requirements.
- Sub-organizations take advantage of all training opportunities to conduct individual, leader, and crew training.
- Support missions/details accomplished with unit integrity to exercise the chain of command and provide individual training opportunities for first line supervisors, as time permits. Unit taskings can be used to reduce the number of permanent special duty personnel within installations and communities.
- Leaves and passes maximized. When appropriate, block leave may be scheduled.
- Routine medical, dental, and administrative appointments coordinated and scheduled with installation support facilities.

Figure 4-6. Green-Amber-Red Time Management System

4-19. Similarly, a *Green-Red* time management system may be more appropriate for some organizations. Although support requirements vary greatly from installation to installation, the time management system can be modified to accommodate these particular situations. The primary purpose of the time management system is to identify and protect prime time training periods for subordinate organizations.

4-20. When the *Green-Red* time management system is used, organizations in Green periods focus training at collective task proficiency, with leader and individual tasks integrated during multiechelon training. These organizations conduct planned training without distraction and external taskings. Organizations in *Red* periods maximize leader and individual self-development opportunities, execute details, other administrative requirements, and allow the maximum number of soldiers to take leave. All administrative and support requirements should be accomplished with unit integrity. The training of an organization in *Red* periods is on small unit, crew leader, and individual proficiency. Figure 4-7 is an example of a Green-Red time management system.

4-21. Likewise, commanders can utilize the *Green-Amber-Red* or *Green-Red* time management system for internal organizational use. For example, it may be feasible for organizations in *Red* periods to meet all support requirements with only a portion of their subordinate units at any given time. In this case, the remaining subordinate units are free to train. A subordinate unit could be assigned an internal *Green* period. In this manner, organizations can optimize small unit, crew, leader, and individual soldier unit training opportunities.

4-22. RC unit commanders can also use either the Green-Amber-Red or Green-Red time management system. Using the *Green-Amber-Red* system: most of the AT period should be *Green* cycle training on collective tasks, most of the IDT periods should be *Amber* and focused on small unit, crew, leader and individual soldier training, with one or two IDT periods being *Red* for mandatory training/administrative requirements. Using the *Green-Red* system: AT and the preponderance of IDT periods should be *Green*, with only one or two IDT periods being Red for mandatory training/administrative requirements. In all training periods DMOSQ training for soldiers who are not school trained and qualified is the first order of business.

➢ ***Green Cycle*** • Training focus primarily on collective tasks with individual and leader tasks integrated during multiechelon training. • Maximum soldier attendance at prime time, mission essential training. • Coincides with availability of major resources and key training facilities or devices. • Administrative and support requirements that keep personnel from participating in training eliminated to the maximum extent possible. • Leaves and passes limited to the minimum essential.
➢ ***Red Cycle*** • Maximize self development. • Diverts the minimum essential number of personnel to perform administrative and support requirements. • Small unit, crew, leader and individual soldier training emphasized. • Sub-organizations take advantage of all training opportunities to conduct individual, leader, and crew training. • Schedule and perform periodic maintenance services. • Support missions/details accomplished with unit integrity to exercise the chain of command and provide opportunities for first line supervisors, as time permits. Unit taskings can be used to reduce the number of permanent special duty personnel within installations and communities. • Provides time for soldier attendance at education and training courses. • Leaves and passes maximized. When appropriate, block leave may be scheduled. • Routine medical, dental, and administrative appointments coordinated and scheduled with installation support facilities.

Figure 4-7. Green-Red Time Management System

TRAINING EVENTS

Do less, do it well, meet the standard...treat every training event as though it were your last. Get to excellence in warfighting.

General Eric Shinseki

4-23. Commanders link training strategies to executable training plans by designing and scheduling training events. During long-range planning, commanders and their staffs make a broad assessment of the number, type, and duration of training events required to accomplish METL training. The event itself is only a tool to achieve and sustain proficiency on the METL within the Band of Excellence. METL proficiency is the objective. In the subsequent development of short-range training plans, senior commanders fully define training events in terms of METL based training objectives, scenarios, resources, and coordinating instructions. Through training events, senior commanders—

- Develop mission-related scenarios.
- Focus the entire organization on several METL tasks.
- Integrate all BOS into coordinated combined arms training.

4-24. Training events are the common building blocks that support an integrated set of METL-related training requirements. Included in long-range training plans, training events form the framework for resource allocation and provide early planning guidance to subordinate commanders and staffs.

4-25. By developing and coordinating training events, the organization is able to bring together the training areas and facilities, TSS products and services, OPFOR, observer/controllers, evaluators, and other resources that create the most realistic and battle focused training. Figure 4-8 shows typical training events.

• Joint Training Exercise (JTX) • Situational Training Exercise (STX) • Command Field Exercise (CFX) • Command Post Exercise (CPX) • Logistic Exercise (LOGEX) • Live Fire Exercise (LFX) • Map Exercise (MAPEX) • CTC Rotations (CTC)	• Combined Training Exercise (CTX) • Tactical Exercise Without Troops (TEWT) • Deployment Exercise (DEPEX) • Combined Arms Live Fire Exercise (CALFEX) • Field Training Exercise (FTX) • Fire Coordination Exercise (FCX) • BCTP/BCBST and other Simulations

Figure 4-8. Types of Training Events

4-26. During planning, senior commanders allocate maximum training time to subordinates. Some large-scale training events, however, must be planned so senior commanders can exercise and integrate all BOS within their wartime organizations. The training value of these large-scale exercises to the entire command is increased when subordinate headquarters participate in developing multiechelon training objectives and scenarios. Subordinate commanders use higher headquarters training guidance, their unit METL, and their unit battle tasks to develop their unit training plan. The next higher commander approves, protects, and resources that plan.

4-27. The Army has increasingly emphasized externally supported training events in which a headquarters senior to the unit being trained provides assistance in the form of detailed planning, additional resources, and evaluation. Support provided by the higher headquarters usually includes a METL derived scenario with associated training and evaluation outlines (T&EO), an OPFOR, observer/controllers, and evaluation support. The Army's CTCs are prime examples of training opportunities that provide combined arms battle focused training that is externally supported. CTCs provide training events based on each participating organization's METL and conducted under realistic and stressful conditions. Externally supported training events can also be conducted at home station to enable units to focus exclusively on the execution of training.

4-28. Organizations can only obtain the full training benefits of externally supported events through carefully planned preparatory training. Therefore, a priority during long-range planning is to develop METL based training programs designed to assist leaders and units in achieving and sustaining

METL task proficiency in the Band of Excellence. METL proficiency is the goal, not the completion of the event.

LIVE, VIRTUAL, AND CONSTRUCTIVE (L-V-C) TRAINING

4-29. Commanders use a mix of live, virtual, and constructive (L-V-C) training to achieve and sustain unit and staff proficiency on selected METL tasks and supporting unit and staff battle tasks within the Band of Excellence. The goal is to train mission essential tasks to standard and sustain a wartime readiness posture. Battalion level units attain and sustain warfighting proficiency and develop soldier fieldcraft primarily through live training. Brigades and higher units rely more on V-C training events to attain and sustain their warfighting proficiency. In general, commanders at battalion level and lower plan and execute standards based in V-C simulations to—

- Prepare for live "in the dirt" training.
- Rehearse selected staff and unit battle tasks and squad, team and crew drills.
- Retrain on selected unit battle tasks, supporting squad, team and crew critical tasks, and leader and individual soldier tasks evaluated as either "P" (needs practice) or "U" (untrained).

4-30. Battalion commanders leverage V-C training events to accelerate junior leader mastery of tasks directly related to developing tactical competence, confidence, and proficiency that support their unit's METL or supporting critical collective tasks. Similarly, battalion and company commanders look to their CSM, 1SGs and key NCO leaders to leverage V-C training events to accelerate junior NCO and soldier mastery of individual tasks directly related to developing their technical competence, confidence, and proficiency that support their small unit, crew, leader, and individual soldier tasks.

4-31. Figure 4-9 provides some of the possible options commanders can use to train soldiers, staffs, leaders, units, and themselves. The commander selects the tools that will result in the unit receiving the best training based on available resources. Virtual and constructive training cannot replace all live training. They can, however, supplement, enhance, and complement live training to sustain unit proficiency within the Band of Excellence.

Several Options: Commanders Select the Right Mix!

	Initial Training	*Proficiency Training*	*Sustainment Training*
Corps/Army Forces/ Joint Task Force	C	C	LVC
Division	C	C	LVC
Brigade	LVC	LVC	LVC
Battalion/Task Force	LVC	L	LV
Company/Battery/Troop	LV	L	LV
Platoon	LV	L	L/V
Crew/Squad	LV	L	L/V
Individual	LV	L/V	L/V

POTENTIAL MIX OF OPTIONS

Live (L)—Training executed in field conditions using tactical equipment enhanced by training aids, devices, simulators, and simulations (TADSS) and tactical engagement simulation to simulate combat conditions.

Virtual (V)—Training executed using computer-generated battlefields in simulators with approximate characteristics of tactical weapon systems and vehicles. Virtual tactical engagement simulation training permits units to maneuver over much larger areas.

Constructive (C)—Training that uses computer models and simulations to exercise the command and staff functions of units from platoon through echelons above corps.

Figure 4-9. Live, Virtual, and Constructive Training Mix

4-32. Battalion task force through division/corps/Army forces/joint forces commanders must sustain their battle staffs' wartime proficiency. Leaders and staffs achieve and sustain proficiency primarily through repetitive execution of battle staff drills to standard. Brigade and battalion/task forces may execute live, virtual, and constructive training. The use of virtual and constructive training provides excellent training opportunities for leader training. The repetitive nature of these tools makes them invaluable in training adaptive leaders. Commanders can run multiple iterations of a task, changing only the conditions, to help mature the decision-making and judgment abilities of subordinate leaders.

4-33. The intent is to train to standard. It is the commander's responsibility to be familiar with all three of these tools and to select the most applicable within allocated L-V-C resources and available training time. The commander, when planning training, must determine the appropriate mix that meets the unit training requirements and objectives. Units may conduct training using L-V-C training, simultaneously.

TRAINING RESOURCES

4-34. The commander uses his assessment of METL and battle tasks to determine the resource priorities for training requirements. During both long-range and short-range planning, constrained resources may require deletion of low-priority training requirements, substitution of less costly training alternatives, or a request for additional resources to execute METL training not resourced, and lower priority training. To the extent possible, commanders confirm resources before publishing training plans. Figure 4-10 lists common sources for information.

- Command operating budget
- Fuel allocation
- Higher headquarters training plans
- Flying hour program
- Training land and range availability
- Force integration documents
- Ammunition authorizations
- Local training area directives
- Availability of TSS products and services, including TADSS

Figure 4-10. Sources of Training Resource Information

4-35. A METL-based events approach to resource planning is used for the allocation of time, facilities, ammunition, funds, fuel products, and other resources. For example, a reasonably close approximation of the future petroleum, oil, and lubricants (POL) (Class III) and repair parts (Class IX) resource requirements (the most significant operations and maintenance costs in a tank battalion) can be calculated for a training event as shown in figure 4-11.

LIN	*System*	*Number Used*	*Miles Traveled*	*Cost Factors*		*System Cost*
				Class IX	*Class III*	
T13305	TANK COMBAT FULL TRACKED: 120 MM GUN M1A2	44	64	218.60	5.87	632,107.52
T92242	TRUCK UTIL ARMT 1-1/4 TON	5	125	0.43	0.06	306.25
T92310	TRUCK UTIL ARMT 1-1/4 TON	5	125	0.43	0.06	306.25
C10990	CARRIER 120 MM MORTAR: SELF-PROPELLED ARMORED	4	60	13.31	0.47	3,307.20
D11538	CARRIER COMMAND POST: LIGHT TRACKED M577	5	50	15.70	0.47	4,042.50
C18234	CARRIER PERSONNEL FULL TRACKED: ARMORED (RISE) M113	10	75	11.65	0.47	9,090.00
T61494	TRUCK UTILITY: CARGO/TROOP CARRIER 1-1/4 TON 4X4 W/E (HMMWV)	21	175	0.43	0.06	1,800.75
T60081	TRUCK CARGO: 4X4 LMTV W/E	7	125	0.77	0.12	778.75
T60149	TRUCK CARGO: 4X4 LMTV W/E W/W	1	125	0.77	0.12	111.25
R50681	M88 RECOVERY VEHICLE	7	75	83.01	1.62	44,430.75
T59278	HEMMT	23	150	1.80	0.34	7,383.00
X40931	TRUCK CARGO: DROP SIDE 5 TON 6X6 W/WINCH W/E	7	175	0.66	0.15	992.25
						$704,656.47

- Calculations derived using FY01 U.S. Army Cost and Economic Analysis Center (USACEAC) cost factors.
- Calculation example: Pieces of Equipment (44) X Miles Traveled (64) X Class IX Cost Factors (218.60) = Class IX Costs
- Pieces of Equipment (44) X Miles Traveled (64) X Class III Cost Factors (5.87) = Class III Costs
- Class IX Costs ($615,577.60) + Class III Costs ($16,529.92) = Total System Costs ($632,107.52)

Figure 4-11. Example Projection of Costs for an active Component Tank Battalion FTX

4-36. The same procedure is followed to determine the costs for each projected training event and totaled into an aggregate training cost for the year. See figure 4-12.

LIN	System	Number Used	Miles Traveled	Cost Factors Class IX	Cost Factors Class III	System Cost ($)
T13305	TANK COMBAT FULL TRACKED: 120 MM GUN M1A2	44	800	218.60	5.87	7,901,340
T92242	TRUCK UTIL ARMT 1-1/4 TON	5	2000	0.43	0.06	4,900
T92310	TRUCK UTIL ARMT 1-1/4 TON	5	2000	0.43	0.06	4,900
C10990	CARRIER 120 MM MORTAR: SELF-PROPELLED ARMORED	4	480	13.31	0.47	26,460
D11538	CARRIER COMMAND POST: LIGHT TRACKED M577	5	375	15.70	0.47	30,320
C18234	CARRIER PERSONNEL FULL TRACKED: ARMORED (RISE) M113	10	545	11.65	0.47	66,050
T61494	TRUCK UTILITY: CARGO/TROOP CARRIER 1-1/4 TON 4X4 W/E (HMMWV)	21	2250	0.43	0.06	23,150
T60081	TRUCK CARGO: 4X4 LMTV W/E	7	2000	0.77	0.12	12,460
T60149	TRUCK CARGO: 4X4 LMTV W/E W/W	1	2000	0.77	0.12	1,780
R50681	M88 RECOVERY VEHICLE	7	675	83.01	1.62	399,880
T59278	HEMMT	23	930	1.80	0.34	45,770
X40931	TRUCK CARGO: DROP SIDE 5 TON 6X6 W/WINCH W/E	7	2000	0.66	0.15	11,340
						$8,528,350

- Calculations derived using FY01 U.S. Army Cost and Economic Analysis Center (USACEAC) cost factors.
- Calculation example: Pieces of Equipment (44) X Miles Traveled (800) X Class IX Cost Factors (218.60) = Class IX Annual Costs
- Pieces of Equipment (44) X Miles Traveled (800) X Class III Cost Factors (5.87) = Class III Annual Costs
- Class IX Annual Costs ($7,694,720.00) + Class III Annual Costs ($206,624.00) = Total System Annual Costs ($7,901,344.00)

Figure 4-12. Example Annual Training Costs for an Active Component Tank Battalion

4-37. There is a relationship between the number of miles or hours that an item of equipment, such as a tank, is operated and the dollars required to purchase the repair parts and POL for that piece of equipment. Funding authority to purchase the projected repair parts, fuel products, and other items necessary to support the training mission is allocated to units based on operating tempo (OPTEMPO). The OPTEMPO of an organization is the average annual miles or hours of operation for its major equipment systems. The total annual training cost of the desired list of training events, as shown in the example at figure 4-12, which represents an OPTEMPO of 800 miles per tank, is then compared with budget projections to determine if the desired training can be fully resourced. If the battalion is not projected to receive sufficient resources to finance the projected list of events, the commander may have to revise the list of events, as illustrated in figure 4-13.

Event	Desired Number of Live Events (Per Year)	Training Event Mix (Reduced Resources)			Total Number of Resourced Events
		Live	Virtual	Constructive	
Battalion CPX	4	0	0	4	4
Battalion CFX[1]	2	2	0	0	2
Battalion FTX	2	1	1	0	2
Company CFX[2]	4	2	2	0	4
Company FTX[3]	4	1	2	0	3
Company LFX[4]	2	2	1	0	3
Platoon FTX[3]	4	1	2	0	3
Platoon LFX[5]	4	2	1	0	3
Estimated Cost	*$8.9M*				*$5.8M*

[1]Battalion CPX: May conduct company level CCTT exercises in conjunction with Battalion CPX (JANUS Battle Staff exercise)
[2]Company CFX: Leader TEWT and CCTT exercises may precede CFX.
[3]Platoon/Company FTX: Normally consists of METL-based STX and one FTX annual external evaluation.
[4]Company LFX: TEWT and CCTT exercises may precede actual LFX.
[5]Platoon LFX: Executed during unit gunnery periods; UCOFT used for preparation and retraining.

Figure 4-13. Revised List of Training Events to Meet Fiscal Constraints

4-38. The Army relies on live FTXs to provide realistic training. Live fire exercises (LFX), STXs, deployment exercises, and battle drills must be conducted under conditions that replicate actual combat as nearly as possible. This is especially true at battalion level and below. Virtual and constructive training cannot replace live training. They can, however, supplement, enhance, and complement live training to sustain unit proficiency within the Band of Excellence. Based on resources available (such as time, ammunition, simulations, and range availability), commanders determine the right mix and frequency of live, virtual, and constructive training to ensure efficient use of allocated training resources. Brigade size and larger organizations normally plan and execute more virtual and constructive training, as shown in figure 4-14.

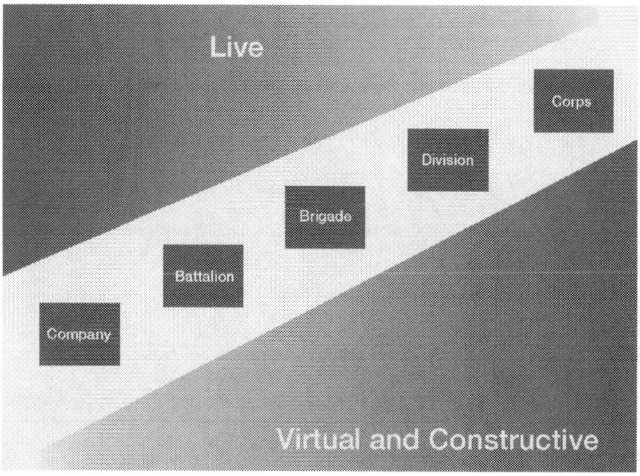

Figure 4-14. Application of Live-Virtual-Constructive Training

4-39. Live, virtual, and constructive training opportunities are integral components of a commander's training strategy to develop competent, confident, and adaptive leaders, battle staffs, and units. A resource analysis allows leaders at all echelons to make training trade-offs, within various budget and program levels, that best support the commander's training strategy.

4-40. The unit may be required to conduct fewer FTXs and LFXs (which require higher densities of equipment and higher resource expenditures) and add a mix of simulation exercises to stay within resource constraints and maintain training proficiency within the Band of Excellence. The commander determines the effect these substitutions will have on attaining desired levels of training proficiency. He then provides this information to the next higher commander who will either provide additional resources or approve the constrained resource plan.

4-41. By summing up fiscal resource projections of subordinate units, commanders at higher echelons are able to estimate resource requirements necessary to support their training strategies. Similar analyses are conducted to estimate ammunition, facilities, and other resources. Upon completion of the trade-off analysis, the commander includes the resulting events and associated resources in the long-range training plan.

4-42. A significant resource consideration in RC planning is the allocation of available training time. Limited training time requires RC commanders to prioritize training requirements. They may have to train fewer tasks so that the Army standard can be attained. RC commanders compensate for lack of training time by carefully distributing requirements over longer periods of time and identifying selected training tasks for execution during postmobilization training. Virtual and constructive training can complement live training because they are time efficient, and can support sustainment of the RC unit in the Band of Excellence.

SHORT-RANGE PLANNING

4-43. Short-range training plans define in greater detail the broad guidance on training events and other activities contained in the long-range training guidance and long-range calendar. They refine the allocation of resources to subordinate organizations and provide a common basis for preparing near-term training plans.

SHORT-RANGE TRAINING GUIDANCE

4-44. Each echelon from division through battalion publishes short-range training guidance that enables the commander and staff to prioritize and refine mission essential training guidance contained in the long-range CTG. Commanders must publish the short-range training guidance with sufficient lead time to ensure subordinate units have time to develop their own short-range training plans. As shown in figure 4-15, the AC division provides quarterly training guidance (QTG) to subordinate commands and installations at least 90 days prior to the start of each quarter. After receiving guidance from higher headquarters, subordinate units down to battalion sequentially publish their QTG. The RC process is conceptually the same as the AC process; except, the guidance normally is published annually as yearly training guidance (YTG) as shown at figure 4-16. Additionally, RC unit commanders are required to develop a postmobilization training plan to complete training to the level organized. This plan should be updated concurrently with the yearly training plan.

Action	*Planning Guidance Publication Date*	*Future Planning Horizon*
Division, separate brigade, regiment, group, or similar level command publishes QTG	3 months prior to start of quarter	3 months
Brigade and group publish QTG	2 months prior to start of quarter	3 months
Battalion, squadron, and separate company publish QTG	6 weeks prior to start of quarter[1]	3 months
Conduct QTB	Prior to start of quarter	3+ months

[1]To allow sufficient time for near-term planning at company level before the start of the quarter.

Figure 4-15. Active Component Short-Range Planning Cycle (Quarterly)

Action	Planning Guidance Publication Date	Future Planning Horizon
Division, separate brigade, regiment, group, or similar level command publishes YTG	6-8 months prior to FY start	1 year
Brigade and separate battalion publish YTG	4-6 months prior to FY start	1 year
Battalion, squadron, and separate company publish YTG	3-4 months prior to FY start	1 year
Conduct YTB	Prior to FY start	1+ years

Figure 4-16. Reserve Component Short-Range Planning Cycle (Annually)

4-45. An important aspect of the quarterly and yearly training guidance is the role of the NCO. Within the framework of the commander's guidance, the CSM/1SG and key NCOs provide planning recommendations on the organization's individual training program. They identify the individual training tasks that must be integrated into collective mission essential tasks during the short-range planning period. Examples of topics normally addressed in QTG and YTG are—

- Commander's assessment of METL proficiency.
- Training priorities and strategy to improve and sustain METL proficiency.
- Combined arms training.
- Organizational inspection program.
- JIM training, as applicable.
- A cross reference of training events and associated METL training objectives.
- Individual training.
- Leader development and leader training.
- Self development.
- Training of trainers and evaluators.
- Training evaluation and feedback.
- Force integration.
- Resource guidance.
- Training management.
- Risk management.

SHORT-RANGE PLANNING CALENDAR

4-46. The short-range planning calendar refines the long-range planning calendar and provides the time lines necessary for small unit leaders to prepare near-term training schedules.

4-47. In preparing a short-range calendar, details are added to further define the major training events contained on the long-range planning calendar. Some examples of these details include—

- The principal daily activities of major training events.
- Home station training conducted in preparation for major training events and evaluations.
- Other mandatory training that supports METL and warfighting such as command inspections as part of the OIP, Army Physical Fitness Test, weapons qualification, or periodic equipment maintenance and services.
- Significant non-training events or activities that must be considered when scheduling training. Examples are national or local holidays and installation support missions.

4-48. The short-range training calendar is coordinated with appropriate installation support agencies to create a common training and support focus between supported and supporting organizations.

TRAINING EVENTS

4-49. Major training events are identified and scheduled during the long-range planning process. Short-range planning refines major training events. A major aspect of designing short-range training events is allocation of time to ensure the planned training is conducted to standard. Detailed information on training events may appear in the organization's short-range training guidance or in separate documents such as exercise directives or memorandums of instruction.

MULTIECHELON TRAINING

4-50. Limited time and other resources do not permit developing sequential training programs in which each echelon from lower to higher is successively trained to reach interim "peaks" in proficiency. Therefore, leaders use a multiechelon training approach to plan training events. Multiechelon training allows simultaneous training and evaluation on any combination of individual and collective tasks at more than one echelon. Multiechelon training is the most efficient and effective way to train and sustain proficiency on mission essential tasks within limited periods of training time.

4-51. Figure 4-17 is an example sequence for a division directed AC battalion task force EXEVAL multiechelon training event. This example depicts mission essential training tasks for each echelon from battalion task force through crew. Various exercise techniques (such as multiple integrated laser engagement system [MILES], battle simulation, and live fire), are used to accomplish the specified training objectives.

Day	Phase	Mission Essential Training Tasks[1]	Battalion Task Force	Company	Platoon/ Squad/ Crew	Evaluation HQ[2]
1	A	Alert/upload basic and operational loads	X	X	X	1st Brigade
2	A	Move to an assembly area and assemble the force	X	X	X	1st Brigade
2	A	Prepare the deliberate defense	X	X	X	1st Brigade
3	B	Continue preparation of the deliberate defense	X	X	X	1st Brigade
4	C	Conduct deliberate defense (FTX) (MILES)	X	X	X	1st Brigade
5 6 7	C	Conduct offensive operations (constructive simulation) • Hasty attack • Deliberate attack • Night attack	X	X (Company HQ Only)	X	2d Brigade
8	C	Conduct hasty attack (LFX)			X	1st Brigade
9	D	Mounted navigation exercise	All Officers			2d Brigade
10	D	Conduct tactical movement (redeploy) and post-operations maintenance	Noncommissioned Officers Control			2d Brigade

[1]Selected individual tasks will be evaluated during each collective training activity.
[2]Division HQ is the exercise control headquarters. Designated brigades provide controllers and evaluators, OPFOR, and range safety personnel.

Figure 4-17. Division Directed Battalion Task Force EXEVAL Multiechelon Training Event

4-52. The designation of control and evaluation organizations is an important aspect of externally supported training exercises. This allows the units performing training to focus on execution of training while other organizations provide the necessary control, evaluation, and administrative support. The 10-day battalion task force EXEVAL training event illustrated in figure 4-17 describes two approaches to multiechelon training (figures 4-18 and 4-19). Multiechelon training occurs when—

- An entire organization focuses on a single METL task. For example, figure 4-18 highlights a number of supporting tasks that a battalion task force performs simultaneously to execute a successful deliberate attack against a prepared enemy defense.
- Different echelons of an organization conduct training on related METL tasks simultaneously. The example at figure 4-19 depicts different echelons training on related tasks during days 5 through 7 of the battalion task force EXEVAL training event at figure 4-17. The battalion task force headquarters and company headquarters participate in a constructive battle simulation while tank platoons concurrently conduct platoon STXs in virtual simulations. Mechanized platoons concurrently conduct squad live fire exercises and crew proficiency training to prepare for the platoon hasty attack LFXs on "Day 8".

Responsibility	*Participants*	*Activity*	*Collective Tasks*	*Leaders/Staff/Individual Tasks*
Task Force Commander	Battalion TF	FTX	• Attack a stationary force • React to chemical attack	***TF CDR:*** Develop attack plan ***S2:*** Develop R&S plan. ***FSO:*** Develop concept of fires. ***CHEMO:*** Develop NBC survey plans.
Team Commander	All assigned or attached personnel	FTX	• Conduct assault of an enemy position • React to chemical attack	Employ direct and/or indirect fires.
Platoon Leader	All assigned or attached personnel	FTX	• Conduct initial breach of mined wire obstacle • Enter/clear trench • React to chemical attack	Analyze terrain; conduct platoon maneuver, plan for use of supporting fires.
Squad Leader	Squad members	FTX	• Enter/clear trench • Knock out a bunker • React to chemical attack	Analyze terrain; conduct squad maneuver; control organic fires.
Individual Soldier	Squad member	FTX	• Enter/clear trench • Knock out a bunker • React to chemical attack	Move as a member of a fire team; engage targets with M16A2, M203 and M249; employ hand grenades.
Note. Leader/Staff/Individual tasks taken from ARTEPs 71-2-MTP, 71-1-MTP and 7-7J-DRILL.				

Figure 4-18. Selected Tasks Planned to be Executed During a Multiechelon Task Force Deliberate Attack

Responsibility	**Participants**	**Activity**	**Tasks**
Task Force Commander	Battalion task force staff and company/team commanders	Constructive simulation	• Attack a stationary force at night.
Mechanized Platoon Leaders	All assigned and attached squad members	Live fire exercise	• Enter/clear a trench • Knock out a bunker
IFV Commanders	Crew Members	Crew proficiency course (CPC)	• IFV Gunnery
Tank Platoon Leaders	All assigned and attached platoon members	Close combat tactical trainer (CCTT)	• Attack by fire • Support by fire

Figure 4-19. Multiechelon Training Exercise Where a Task Force Executes Related Tasks While Simultaneously Conducting Different Training Activities

4-53. Figure 4-20 depicts a multiechelon training concept for an RC division annual training period. It addresses some RC unique training considerations such as the use of the CONUSA, division (training support), and AC support of RC training.

Event	Description
Wartime Mission-Related CPX	All commanders and staffs from division through battalion participate in an exercise that thoroughly rehearses wartime operations plans. Division (training support) provides controllers, operates the battle board, and simulates the company level chain of command. The CONUSA provides personnel for a corps headquarters response cell and assists in evaluation.
Company and Platoon STX	As more senior commanders are participating in the CPX, companies negotiate a series of METL-related STXs. For example, an RC maneuver platoon is required to cross an LD at a specific time, react to an enemy ambush, clear an obstacle, conduct a hasty attack, and defend against a counter attack. The RC unit would perform the STX, participate in detailed after-action reviews, and renegotiate the course until the Army standard on each training task was achieved. Similar STXs are established for all of the combat arms, combat support, and combat service support organizations in the division.

Figure 4-20. Multiechelon Training for Reserve Component Division Annual Training

4-54. Larger scale training events also provide an opportunity for valuable individual, crew, battle staff, and small unit training. These exercises can result in unproductive training for soldiers at lower echelons unless senior leaders plan multiechelon training down to the smallest participating units. This is the best method to maintain battle focus on the large unit METL tasks as well as on supporting collective and individual battle tasks for even the smallest participating units.

TRAINING RESOURCES

4-55. In short-range planning, commanders allocate training resources to subordinate organizations for specific training activities. As required, adjustments are made from the initial resource projections contained in long-range plans. The key requirement for division and brigade commanders is to coordinate short-range training plans with the various resource processes that support training. Examples of these processes are Program Budget Advisory Committee (PBAC) meetings, ammunition forecasts, and training area and facility scheduling conferences. A significant resource to assist the commander in planning training is the TSS. The TSS is a collection of resources that supports training and leverages available technology to replicate combat conditions and enhance training. Examples of TSS training support products are—

- Facilities such as ranges, training areas, firing points, urban training sites, digital training facilities, and mission support and training facilities.
- Training products such as MTP, training support packages, multimedia products, and distance learning through electronically stored and delivered course content and programs of instruction.

- TADSS such as tactical engagement simulations, instrumentation at the CTC and home station, embedded training capabilities, MILES, and warfighter simulation.
- Training services such as the Center for Army Lessons Learned, proponent schools, installation support, and CTC.

Training support products and their applications to training will be discussed in detail in FM 7-1 (projected).

TRAIN THE TRAINERS

4-56. Training the trainers is a critical step in preparation for training. The leaders, trainers, and evaluators involved in any training event must know, understand, and be proficient on the specified tasks. Leaders, trainers, and evaluators must be trained to standard if the training event is to be done to standard.

4-57. In addition to leader training, specific trainer training must also be identified and planned. All leaders are trainers, but all trainers are not necessarily leaders. A specialist or subject matter expert may be necessary to conduct the instruction for a particular collective or individual task. It is essential that these trainers be allocated sufficient time to prepare the specified training.

4-58. An overlap in training responsibilities, frequently overlooked by leaders when planning, is the case where a subordinate leader is the primary trainer as well as the leader of an element undergoing a collective training event. Senior leaders must consciously allocate sufficient time for subordinates to prepare for these responsibilities.

4-59. Training leaders, trainers, and evaluators to standard supports, enhances, and enables collective training when properly planned and conducted before the training event. Commanders must plan, resource, and ensure timely accomplishment of trainer training.

SHORT-RANGE TRAINING BRIEFINGS

4-60. The short-range training briefing is a conference conducted by senior commanders to review and approve the training plans of subordinate units. It is conducted before the time period addressed in the QTG or YTG. AC units conduct QTB. RC units conduct YTB.

4-61. Division commanders receive the short-range training briefing from subordinate brigades and all battalions in the division. The brigade commander and CSM personally present the overview of the brigade training plan; battalion commanders and CSMs present detailed briefings of their training plans. All habitually associated commanders participate in preparing and conducting the training briefing.

4-62. Training briefings produce a contract between the senior commander and each subordinate commander. As a result of this contract, the senior commander agrees to provide resources, including time, and protect the subordinate unit from unprogrammed taskings. The subordinate commander agrees to execute the approved training plan and conduct training to standard. This shared responsibility helps maintain priorities, achieve unity of effort, and synchronize actions to achieve quality training and efficient resourcing. The QTB or YTB, as appropriate, is the forum where contracts for that training period are discussed and confirmed. Training guidance flows from the top-down and requirements for planning and execution of tasks flow from the bottom-up.

4-63. The training briefing is a highlight of the senior commander's leader development program. It provides the commander an opportunity to coach and teach subordinates on the fine points of his philosophy and strategies in all aspects of warfighting, to include doctrine, training, force integration, and leader development. It enables subordinate commanders, some of whom may be new to the organization, to gain a better understanding of how their mission essential training relates to the battle focused training programs of their senior commanders and peers.

4-64. The senior commander specifies the format and content of the briefing in the QTG or YTG. However, the briefing guidance should be flexible enough to provide subordinate commanders and CSMs the latitude to highlight their initiatives and priorities. Units should refrain from discussing readiness issues not directly related to training. Such statistical, logistical, manning, or other management data is more appropriate to other readiness review forums and distracts from the overall training focus of the QTB or YTB.

4-65. During the training briefing, the subordinate commanders, as a minimum, usually address the following specific areas—

- Brief training that was planned and briefed at previous QTB or YTB, but was not conducted and why.
- The organization's METL and assessment of proficiency levels.
- A discussion of the unit's training focus and objectives for the upcoming training period.
- A presentation of the organization's short-range planning calendar.
- A description of upcoming training events.
- Officer leader development program with emphasis on warfighting skill development.
- Self development.
- Risk management.
- Plans for preparing trainers and evaluators.
- Force integration plans for the upcoming period.
- Resource allocation.

4-66. Each CSM normally follows the commander's presentation. The CSM provides an analysis of the organization's individual training proficiency and discusses the organization's planned individual training and education. Example discussion topics include—

- Individual training proficiency feedback received concerning previous short-range planning period.
- An assessment of the organization's current individual training proficiency.

- Individual training events planned during the upcoming short-range planning period and strategy to prepare soldiers for these evaluations.
- A description of METL derived individual tasks to be integrated with upcoming collective mission essential tasks.
- Marksmanship and physical fitness programs.
- NCO leader development program with emphasis on warfighting skill development.
- Self development.
- NCO/enlisted schools.

NEAR-TERM PLANNING

4-67. Near-term planning is primarily conducted at battalion and subordinate command levels. It is conducted to—

- Schedule and execute training objectives specified in the short-range training plan to the Army standard.
- Provide specific guidance to trainers.
- Make final coordination for the allocation of resources to be used in training.
- Complete final coordination with other units that will participate in training as part of the task organizations.
- Prepare detailed training schedules.

4-68. Near-term planning covers a six- to eight-week period prior to the conduct of training for AC units (figure 4-21) and a four-month period prior to training for RC units (figure 4-22). Formal near-term planning culminates when the unit publishes its training schedule.

Action	*Planning Guidance Publication Date*	*Future Planning Horizon*
Battalion training meetings and subsequent draft training schedules[1]	6-8 weeks prior to execution	6-8 weeks
Battalion publishes training schedules[2]	4-6 weeks prior to execution	4-6 weeks
[1]Training schedules are developed at company level and approved by battalion commanders.		
[2]Training schedules are typed and reproduced at battalion level.		

Figure 4-21. Active Component Near-Term Planning Cycle (Weekly)

Action	Planning Guidance Publication Date	Future Planning Horizon
Battalion training meetings and subsequent draft training schedules[1]	4 months prior to execution	4 months
Battalion publishes training schedules	3 months prior to execution	3 months
[1]Training schedules are developed at company level and approved by battalion commanders		

Figure 4-22. Reserve Component Near-Term Planning Cycle (Monthly)

TRAINING MEETINGS

4-69. Training meetings are the key to near-term planning. Training meetings create the bottom-up flow of information regarding specific training proficiency needs of the small unit, battle staff, and individual soldier. Normally platoons, companies, and battalions conduct weekly training meetings. At battalion level, training meetings primarily cover training management issues. At company and platoon level, they are directly concerned with the specifics of training execution and must include pre-execution checks. During training meetings, nothing is discussed but training. All key leaders of the unit must attend.

TRAINING SCHEDULES

4-70. Near-term planning results in detailed training schedules. Training is considered "locked in" when the battalion commander signs the training schedule. At a minimum, it should—

- Specify when training starts and where it takes place.
- Allocate adequate time for scheduled training and additional training as required to correct anticipated deficiencies.
- Specify individual, leader, and collective tasks to be trained.
- Provide concurrent training topics that will efficiently use available training time.
- Specify who conducts the training and who evaluates the training.
- Provide administrative information concerning uniform, weapons, equipment, references, and safety precautions.

4-71. Senior commanders establish policies to minimize changes to the training schedule. Training is locked in when training schedules are published. Command responsibility is established as follows—

- The company commander drafts the training schedule.
- The battalion commander approves and signs the schedule and provides necessary administrative support.
- The brigade commander reviews each training schedule published in his command.
- The division commander reviews selected training schedules in detail and the complete list of organization wide training highlights developed by the division staff.

4-72. Senior commanders provide feedback to subordinates on training schedule quality and subsequently attend as much training as possible to ensure that mission essential tasks are accomplished to standard.

CS AND CSS TRAINING

4-73. CS and CSS units support combined arms unit training every day through execution of core warfighting functional tasks. Combat arms unit commanders recognize their units cannot conduct combined arms training without their task organized CS and CSS units. For example, combat arms unit commanders recognize their units cannot train without operational equipment, fuel, rations, water, and other supplies and services provided by their supporting CSS units. CS and CSS unit commanders integrate their unit training plans with their supported combat arms units. CS and CSS units daily perform their core warfighting functional tasks, at the section, team, and individual technical MOS level. For example, maintenance support teams routinely perform organizational and direct support automotive, turret, armament, and communications-electronic maintenance and periodic services, as well as provide repair parts support to their supported combat arms units. Certain low density technical MOSs pose a particular training challenge because these soldiers may not be able to perform some of their individual technical MOS tasks while their organization is in garrison. CS and CSS unit commanders look for opportunities elsewhere on the installation to train these soldiers on their individual technical MOS tasks.

4-74. CS and CSS unit commanders ensure training exercises are designed in such a way as to provide opportunities to train on CS and CSS company and battalion level METL tasks. For example, a corps support group (CSG) commander may design an exercise that provides an opportunity for a subordinate engineer battalion (combat heavy), a quartermaster company (water supply) (direct support/general support [DS/GS]) and a quartermaster tactical water distribution team (Hoseline) to practice selected wartime METL tasks while participating in a support operations training exercise. In this exercise, these units provide water supply and distribution, and restore vital infrastructure to a host country devastated by a natural or man-made disaster. Figure 4-23 highlights a number of supporting tasks, at different echelons, that an engineer battalion (combat heavy) performs during such a support operation exercise. All these tasks support a single wartime METL task of construct/repair water distribution system.

Responsibility	Participants	Activity	Collective Tasks	Leader/Individual Tasks
Battalion Commander (CBT) (HVY)	Battalion Staff	FTX	Prepare construction estimates Site-adapt a standard construction design Control construction operations	**BN CDR:** Determine events in a construction project **S2:** Plan/Direct engineer intelligence collection **S3:** Schedule work in a construction project
Company Commander (CBT)(HVY)	All assigned or attached personnel	FTX	Prepare a water storage and distribution site	Design drainage system
Platoon Leader (Gen Const PLT)	All assigned or attached personnel	FTX	Construct/repair a water distribution system Construct/repair sewage system	Read construction prints Prepare a bill of materials
Squad Leader (Gen Const PLT)	Squad members	FTX	Conduct excavation operations	Supervise installation of plumbing system
Individual Soldiers	Squad members	FTX	Conduct excavation operations	Backfill with scoop loader Operate excavation equipment
Note Tasks taken from ARTEPs 5-415-66-MTP, 5-417-35-MTP, 5-417-17 and 5-417-17-MTP.				

Figure 4-23. Selected Tasks Executed During a Multiechelon EAD/EAC Engineer FTX

4-75. Figure 4-24 shows different echelons of a quartermaster company (water supply) (DS/GS) and quartermaster tactical water distribution team (Hoseline) conducting training on a single METL task of "Provide water supply and distribution support." This METL task is executed under the conditions of a support operation.

Responsibility	Participants	Activity	Collective Tasks	Leader/Individual Tasks
QM Company (Water Supply) (DS/GS) Commander and Quartermaster Tactical Water Distribution Team (Hoseline) Detachment Commander	All assigned or attached personnel	STX	Perform DS water issue and GS water distribution operations Coordinate company water support operations	**Company/Detachment Commander:** Review mission water requirements to meet total daily water distribution/storage needs of supported force **Supply Control Section:** Coordinate water production, water storage and DS water issue/GS water distribution
Platoon Leader	All assigned or attached personnel	STX	Perform DS water issue and GS water distribution operations. Plan water operations	Set up distribution schedule Plan water storage, issue and distribution operations
Section Leader/Team Leader	Section members	STX	Perform DS water issue and GS water distribution operations Conduct water quality analysis program	Supervise DS water issue and GS water distribution operations Supervise water quality analysis program
Individual Soldier	Section member	STX	Perform DS water issue and GS water distribution Conduct water quality analysis program	Operate/perform PMCS on the 600-GPM Diesel water pump Maintain, assemble and disassemble the tactical water distribution system (TWDS) hoseline Conduct H_2O analysis testing
Note. Collective/Leader and Individual tasks taken from ARTEP 10-468-30-MTP and STP 10-77W14-SM-TG				

Figure 4-24. Selected Tasks Executed During a Multiechelon EAD/EAC Water Supply and Distribution STX

GARRISON TRAINING

4-76. Garrison commanders' training plans incorporate mobilization, post-mobilization, deployment, redeployment, and demobilization requirements. Garrison commanders plan and schedule periodic mobilization exercises (MOBEXs), emergency deployment readiness exercises (EDREs), and other contingency plan exercises to sustain proficiency on Title X related tasks outlined in current Army and MACOM regulations. Garrison commanders coordinate their training plans with their supported corps, divisional, and tenant organizations. Garrisons routinely support scheduled unit training deployments and exercise certain deployment tasks such as "operating departure/arrival airfield control groups and seaports of embarkation and debarkation."

Chapter 5

Execution

Only through high training requirements, rigidly enforced, can low casualty rates be possible. Only well armed and equipped, adequately trained and efficiently led forces can expect victory in future combat.

General Matthew B. Ridgway

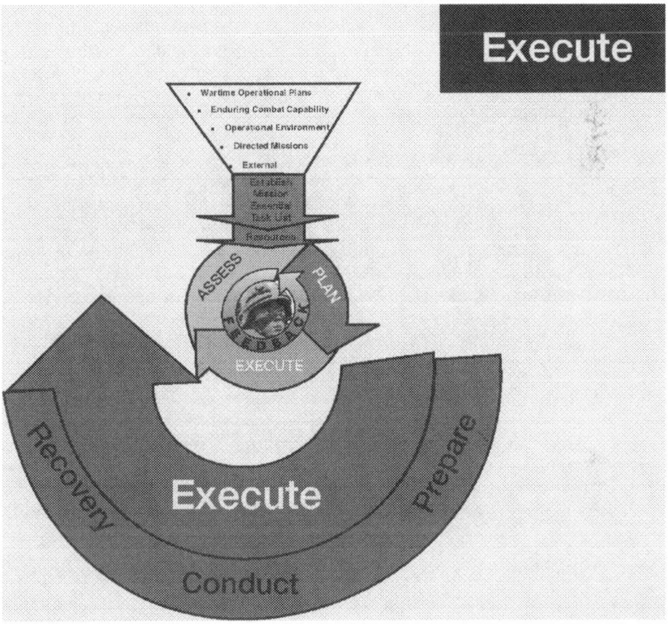

EXECUTION OF TRAINING

5-1. All good training, regardless of the specific collective, leader, and individual tasks being executed, must comply with certain common requirements. These include adequate preparation, effective presentation and practice, and thorough evaluation. (Evaluation is discussed in chapter 6.) The execution of training includes preparation for training, conduct of training, and recovery from training (figure 5-1).

5-2. The training execution process is applicable at all echelons, from a high level staff participating in a joint training exercise to a first line leader's individual training of his team.

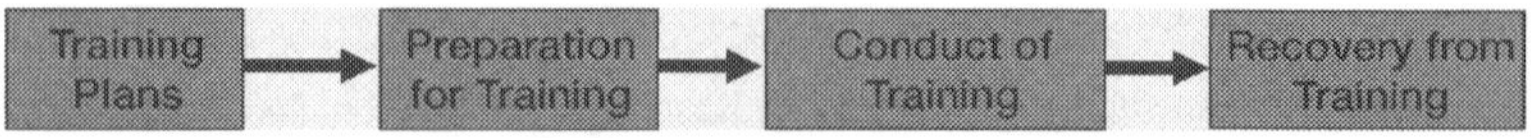

Figure 5-1. Training Execution Process

PREPARATION FOR TRAINING

5-3. As discussed in chapter 4, formal near-term planning for training culminates with the publication of the unit training schedule. Informal planning, detailed coordination, and preparation for executing the training continue until the training is performed. Commanders and other trainers use training meetings to assign responsibility for preparation of all scheduled training. Preparation for training includes selecting tasks to be trained, planning the conduct of the training, training the trainers, reconnaissance of the site, issuing the training execution plan, and conducting rehearsals and pre-execution checks. Pre-execution checks are preliminary actions commanders and trainers use to identify responsibility for these and other training support tasks. They are used to monitor preparation activities and to follow-up to ensure planned training is conducted to standard. Pre-execution checks are a critical portion of any training meeting. During preparation for training, battalion and company commanders identify and eliminate potential training distracters that develop within their organizations. They also stress personnel accountability to ensure maximum attendance at training.

5-4. Subordinate leaders, as a result of the bottom-up feed from internal training meetings, identify and select the collective, leader, and individual tasks necessary to support the identified training objectives. Commanders develop the tentative plan to include requirements for preparatory training, concurrent training, and training resources. At a minimum, the training plan should include confirmation of training areas and locations, training ammunition allocations, training simulations and simulators availability, transportation requirements, soldier support items, a risk management analysis, assignment of responsibility for the training, designation of trainers responsible for approved training, and final coordination. The time and other necessary resources for retraining must also be an integral part of the original training plan.

5-5. Leaders, trainers, evaluators, observer/controllers, and OPFOR are identified, trained to standard, and rehearsed prior to the conduct of the training. Leaders and trainers are coached on how to train, given time to prepare, and rehearsed so that training will be challenging and doctrinally correct. Commanders ensure that trainers and evaluators are not only tactically and technically competent on their training tasks, but also understand how the training relates to the organization's METL. Properly prepared trainers, evaluators, and leaders project confidence and enthusiasm to those being trained. Trainer and leader training is a critical event in the preparation phase of training. These individuals must demonstrate proficiency on the selected tasks prior to the conduct of training.

5-6. Commanders, with their subordinate leaders and trainers, conduct site reconnaissance; identify additional training support requirements; and refine and issue the training execution plan. The training plan should identify all elements necessary to ensure the conduct of training to standard. Rehearsals are essential to the execution of good training. Realistic, standards based performance oriented training requires rehearsals for trainers, support personnel, evaluators, observer/controllers, and OPFOR.

5-7. Preparing for training in RC organizations can require complex pre-execution checks. RC trainers must often conduct detailed coordination to obtain equipment, TSS products, and ammunition from distant locations. In addition, RC pre-execution checks may be required to coordinate AC assistance from the numbered CONUSA, Divisions (Training Support), and directed training affiliations.

CONDUCT OF TRAINING

5-8. Ideally, training is executed using the crawl-walk-run approach. This allows and promotes an objective, standards-based approach to training. Training starts at the basic level. Crawl events are relatively simple to conduct and require minimum support from the unit. After the crawl stage, training becomes incrementally more difficult, requiring more resources from the unit and home station, and increasing the level of realism. At the run stage, the level of difficulty for the training event intensifies. Run stage training requires optimum resources and ideally approaches the level of realism expected in combat. Progression from the walk to the run stage for a particular task may occur during a one-day training exercise or may require a succession of training periods over time. Achievement of the Army standard determines progression between stages.

5-9. In crawl-walk-run training, the tasks and the standards remain the same, however, the conditions under which they are trained change. Commanders may change the conditions for example, by increasing the difficulty of the conditions under which the task is being performed, increasing the tempo of the task training, increasing the number of tasks being trained, or by increasing the number of personnel involved in the training. Whichever approach is used, it is important that all leaders and soldiers involved understand which stage they are currently training and understand the Army standard.

Example

The crawl-walk-run approach occurs in the execution of a mechanized infantry platoon executing an "Assault an Objective" STX. In the crawl stage, the platoon conducts a dismounted rehearsal of the assault. In the walk stage, the platoon conducts a fully mechanized rehearsal of the assault to include consolidation and reorganization. In the run stage, the platoon executes several iterations of the assault against an OPFOR. Some iterations are conducted under NBC conditions, and some during periods of limited visibility. In each iteration of the assault, the platoon strives to achieve the tactical objective to the standard described in the T&EO for "assault an objective."

5-10. An AAR is conducted immediately after training and may indicate that additional training is needed. Any task that was not conducted to standard should be retrained. Retraining should be conducted at the earliest opportunity. Commanders should program time and other resources for retraining as an integral part of their long, short-, and near-term training planning cycle. Training is incomplete until the task is trained to standard. Soldiers will remember the standard enforced, not the one discussed.

5-11. Commanders employ this same approach using virtual and constructive simulations as a means to train battle staffs and subordinate organizations. Commanders strive to attain and sustain a level of proficiency within the Band of Excellence.

5-12. Trainers use the appropriate combination of demonstrations, conferences, discussions, and practice activities to present training. Using the crawl-walk-run approach, they inform individuals being trained of the training objectives (tasks, conditions, and standards) and applicable evaluation methods. They immediately follow presentation with practice to convert information into usable individual and collective skills. The amount of detail included in practice depends on experience levels. If individuals or organizations are receiving initial training on a mission essential task, trainers emphasize the basic conditions. If those receiving the instruction are receiving sustainment training on a task, trainers raise the level of detail and realism until the conditions replicate the wartime environment as closely as possible. Trainers challenge those with considerable experience to perform multiple training tasks within a given training scenario. Properly presented and executed training is realistic, safe, accurate, well structured, efficient, and effective:

- **Realistic** training requires organizations to train the way they will fight or support within all dimensions of the battlefield/space. Realistic training includes all available elements of combined arms teams and, as appropriate, joint, multinational, and interagency teams. It optimizes the use of TSS products to replicate the stresses, sounds, and conditions of combat.
- **Safe** training is the predictable result of performing to established tactical and technical standards. Through the risk management process, leaders at all echelons ensure that safety requirements are integral and not add-on considerations to all aspects of planning, executing, and evaluating training.
- **Accurate** training complies with Army operational and training doctrine and is technically correct. Field manuals, MTPs, battle drills, and other training publications provide factual information to trainers to facilitate conduct of training, coach subordinate trainers, and evaluate training results.
- **Well-structured** training contains a mixture of initial and sustainment training. It also consists of a mix of individual and leader tasks that are integrated into METL collective tasks. Soldiers and leaders increase proficiency in individual tasks while training on collective mission essential tasks.
- **Efficient** training ensures that training resources are expended properly. Efficiently executed training makes full use of every participant's time. Commanders monitor physical and financial resource execution through PBACs, range conferences, and similar forums. They use the feedback received during these forums to adjust resources within their commands to sustain METL proficiency within the Band of Excellence. Continuing advances in training technology enhance the commander's ability to hone warfighting skills and are increasingly required to balance constraints to training, such as environmental protection considerations and availability of training areas and ranges. Similarly, TSS products and services, such as TADSS, not only

provide a means for initial and sustainment training on warfighting fundamentals, but also provide relatively inexpensive preparation for resource intensive training events. Although TSS products provide excellent virtual and constructive training supplements, there is no substitute for the more robust experience of live training.

- **Effective** training builds proficiency, teamwork, confidence, and cohesiveness. Effective training is competitive. Although individuals and organizations may sometimes compete against one another, they should always compete to achieve the prescribed standard. If they do not initially achieve the standard, trainers take corrective actions so that the proper performance level results. Additional considerations for conducting effective training are—
  - **Training and Evaluation Outline (T&EO).** Effective collective, leader, and individual training are guided by the use of T&EOs. The T&EO provides summary information concerning collective training objectives as well as individual and leader training tasks that support the collective training objectives. They also provide information concerning resource requirements and evaluation standards applicable to a training situation. The principal source documents for T&EOs are MTPs and other soldier training publications. Since the conditions in these publications can vary, trainers adjust and supplement T&EO conditions to conform to the METT-TC of the organization's operational plans.
  - **Individual Training.** Fundamental to the adaptability of the force is the maintenance of individual skills yielding technically and tactically competent soldiers who are confident in their abilities. The individual soldier is the heart of any unit's ability to conduct its mission. The ability to perform individual/leader skills to standard is founded in the institutional training base, but it is honed and maintained by effective, periodic repetition of tasks. Training devices, simulators, and web-based training can be used to facilitate the training of individual tasks.
  - **Leader Training.** Leaders spend virtually all available training time supervising the training of subordinates. Often, they do not increase their own understanding of how to fight as combat or support leaders. Therefore, senior commanders view leader training as a continuous process that encompasses more than periodic officer and NCO professional development classes. Senior commanders establish a positive training environment that encourages subordinates to become adaptive leaders capable of independent thinking on the move, and of timely decision making based on broad, effects-based intent guidance, mission orders, and a shared vision of the battlefield. Growing and maturing leaders is a vital part of an effective training program. Leader training, when properly conducted, produces competent, confident, adaptable leaders, and ultimately produces soldiers who are confident in the abilities of their leaders.
  - **Battle Rosters.** Battle rosters are maintained at battalion level and below to track key training information on selected mission essential systems (such as tanks, howitzers, automated command and control systems, forklifts, etc.). They track such pertinent training data as crew stability and manning levels, and qualification status. A key aspect of battle rosters is the designation of qualified back-up operators or crewmembers assigned in other positions in the organization. During the execution of training, battle rostered crewmembers train with their designated crews

at available opportunities. Commanders must discipline the battle roster system.

- **Battle Staff Training.** Battle staff training develops and sustains planning, coordination, execution, and other staff functions related to wartime mission requirements. Battle staff training objectives are derived from the staff METL. Commanders train battle staffs primarily through a mix of constructive and virtual simulations. They maximize the use of information technology systems to enhance leader skills and to develop the adaptiveness necessary to leverage developing information technology. Battle staffs train to integrate and coordinate the BOS internally within their own headquarters, horizontally with other staffs at the same organizational level, and vertically with higher and subordinate organizational staffs. The result of this training produces commanders and staffs capable of synchronizing the BOS across the full spectrum of operations. A well-trained battle staff is a combat multiplier.
- **JIM Training.** JIM training requires different considerations. Joint training is conducted using approved joint doctrine and TTPs, and must be consistent with assigned joint missions and priorities. When assigned as a JFC, Army commanders establish joint training objectives and plans, execute and evaluate joint training, and assess training proficiency. Multinational training optimizes contributions of member forces by matching their missions with their capabilities, and uses available training assistance programs. Major FTXs can be complemented with the use of simulations to enhance multinational training. Joint training publications are available to multinational partners. Interagency training is of growing importance. When Army missions will likely involve U.S. government agencies, if possible, training should be conducted with the organizations and people likely to be involved in the assigned mission. JIM training is as rigorous as any other training the Army conducts. This training also results in mutual appreciation for other capabilities, and the development of valuable personal and professional relationships among those who will operate together.

RECOVERY FROM TRAINING

5-13. The recovery process is an extension of training and, once completed, it signifies the end of the training event. At a minimum, recovery includes conduct of maintenance training, turn-in of training support items, and the conduct of AARs that review the overall effectiveness of the training.

5-14. Maintenance training is the conduct of post operations preventive maintenance checks and services, accountability of organizational and individual equipment, and final inspections. Class IV, Class V, TADSS and other support items are maintained, accounted for, and turned-in. Training sites and facilities are closed out.

5-15. AARs conducted during recovery focus on collective, leader, and individual task performance, and on the planning, preparation and conduct of the training. Unit AARs focus on individual and collective task performance, and identify shortcomings and the training required to correct deficiencies. AARs with leaders focus on tactical judgment. These AARs contribute to leader learning and provide opportunities for leader development. AARs with trainers, evaluators, observer/controllers, and OPFOR provide additional opportunities for leader development.

5-16. The AARs conducted during recovery along with the AARs that took place during the conduct of training enhance future training . They provide the feedback that contributes to the development of training plans to correct identified deficiencies. Finally, these AARs contribute to the commander's overall evaluation of training effectiveness and unit assessment. However, they are not in themselves the end state of recovery. Recovery from training is complete when the unit is again prepared to conduct its assigned mission.

ROLE OF COMMANDERS AND SENIOR LEADERS

5-17. Although planning for training is relatively centralized to align training priorities at all echelons of an organization, the execution of training is decentralized. Decentralization tailors training execution to available resources and promotes bottom-up communication of unique wartime mission related strengths and weaknesses of each individual, leader, and unit.

5-18. Senior commanders must personally observe and evaluate the execution of training at all echelons to the maximum extent possible. From their observations of training and other feedback, they provide guidance and direct changes that lead to improved training and increased readiness.

5-19. By personally visiting training, senior commanders communicate to subordinate units and leaders the paramount importance of training. In addition to observing and evaluating the training of their headquarters and immediate subordinate commands, senior commanders also observe and evaluate the quality of training at all echelons down to the lowest levels of the organization. They receive feedback from subordinate leaders and soldiers during training visits. Through feedback, senior commanders identify and resolve systemic problems in planning, leadership, management, support, and other functions.

5-20. The most beneficial senior commander and staff visits to training are unannounced or short notice. They observe training as experienced by soldiers and prevent excessive visitor preparation by subordinate organizations (this, in itself, can become a training distracter). Senior commanders assign coordination of training support for subordinate units as a priority requirement for organizational staffs. Training support and coordination of training resources are key to successful training execution. Senior commanders check the adequacy of external training support during every training visit and require prompt and effective corrective action to resolve support deficiencies.

ROLE OF NONCOMMISSIONED OFFICERS

5-21. The difference in our Army and every other army in the world is that we have a proud, professional NCO Corps that takes pride in, and accepts responsibility for, the care and individual training of soldiers. CSM/1SG and key NCOs select and train specific individual tasks that support the units' collective mission essential tasks. NCOs are indispensable throughout the training process. Commanders approve the tasks selected and supervise and evaluate training with the officers and NCOs throughout the training execution process.

5-22. NCOs are responsible for individual, crew, and small team training. They continue the soldierization process of newly assigned enlisted soldiers and begin their professional development. In units, individual skill training is presented by the first-line leader, and not presented to large numbers of soldiers by committee. The first-line leader is responsible to train individual tasks to soldiers in their squads, crews, teams, and equivalent small units. The first-line leader and senior NCOs emphasize standards-based, performance-oriented training to ensure soldiers achieve the Army standard. NCO leaders conduct cross training to ensure critical wartime skills within the unit. The CSMs, 1SG, and other senior NCOs coach junior NCOs to master a wide range of individual tasks. Commanders allocate training time for NCOs to conduct individual training and require that individual tasks be included in all collective METL training. NCOs are responsible for conducting individual training to standard and must be able to explain how individual task training relates to the collective mission essential tasks.

5-23. Individual, crew, and small team tasks to be trained are based on the small unit leader's evaluation of training deficiencies. These tasks are input as the NCO's bottom-up feed at the weekly training meeting, approved by the commander, and incorporated into the unit training plans and subsequent training schedules. NCO leaders plan, prepare, rehearse, execute, and conduct AARs for the approved training and provide feedback during weekly training meetings. Commanders may, as required, approve the conduct of training that may not have a strictly tactical focus but sustains soldier readiness. For example, low-density occupational specialty soldiers may be consolidated periodically for training under the senior functional NCO to sustain proficiency.

Chapter 6

Assessment

The best form of "welfare" for troops is first class training, for this saves unnecessary casualties.

Field Marshal Erwin Rommel

Assessment is the commander's responsibility. It is the commander's judgment of the organization's ability to accomplish its wartime operational mission.

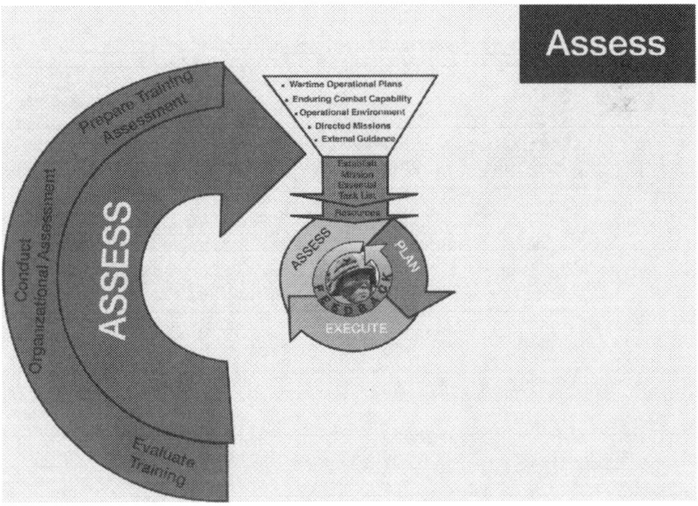

ASSESSMENT

6-1. Assessment is the commander's responsibility. It is the commander's judgment of the organization's ability to accomplish its wartime operational mission. Assessment is a continuous process that includes evaluating training, conducting an organizational assessment, and preparing a training assessment. The commander uses his experience, feedback from training evaluations, and other evaluations and reports to arrive at his assessment. Assessment is both the end and the beginning of the training management cycle.

6-2. Training assessment is more than just training evaluation, and encompasses a wide variety of inputs. Assessments include such diverse systems as training, force integration, logistics, and personnel. They provide the link between the unit's performance and the Army standard. Evaluation of training is, however, a major component of assessment. Training evaluations provide the commander with feedback on the demonstrated training proficiency of soldiers, leaders, battle staffs, and units. Commanders cannot personally observe all training in their organization and, therefore, gather feedback from their senior staff officers and NCOs.

ORGANIZATIONAL ASSESSMENT

6-3. Battalion and higher echelon commanders must be concerned with broader concepts. Accordingly, they perform organizational assessments that aggregate a large number of evaluations. These commanders establish an organizational assessment program that—

- Fixes responsibility within the staff and subordinate units for gathering and analyzing evaluation data and preparing recommendations.
- Concentrates on the effectiveness of leader and organization training.
- Utilizes the CSM and other senior NCOs to gather feedback on the individual, crew, and team training.
- Allows the senior commander to monitor outcomes and take action to reshape priorities, policies, or plans to overcome assessed weaknesses and sustain demonstrated strengths.

6-4. CTC take-home packages are an excellent source of feedback to include in an organizational assessment. These packages consist of videotapes and written documentation of AARs, a report of unit strengths and weaknesses as noted by the observer/controllers, and recommendations for future home station training. Some important sources of feedback for the senior commander's assessment of the organization's ability to accomplish wartime missions are listed in figure 6-1.

- ❐ Personal observations of training.
- ❐ Assessment and feedback from higher headquarters.
- ❐ Staff visit reports.
- ❐ Unit status reports.
- ❐ Training briefings.
- ❐ Local ARTEP evaluations and CTC take-home packages.
- ❐ AARs from FTX, gunnery periods, or other major training exercises.
- ❐ AT reports.
- ❐ CTT results (component of ITEP).
- ❐ UCOFT/MCOFT results.
- ❐ AAR generated reports from training activities.
- ❐ EDRE reports.
- ❐ Maintenance and logistical evaluations and technical inspection results.
- ❐ IG general and special inspections.
- ❐ Commander's Organizational Inspection Program.
- ❐ Force integration reports and feedback.
- ❐ Army Audit Agency reports.
- ❐ APFT scores.
- ❐ Weapon qualification records.
- ❐ Division (training support) assistance input.
- ❐ CTC take-home packages.

FIGURE 6-1. Sources of Feedback for Organizational Assessments

EVALUATIONS

6-5. Evaluations can be informal, formal, internal, external, or any combination, thereof—

- **Informal evaluations** take place when a leader conducts training with his unit, for example when a squad leader trains his squad to assault an objective. Another example would be whenever a leader visits ongoing training, for instance when a battalion commander observes company training. This type of evaluation provides real time feedback on the training environment and the proficiency resulting from training.
- **Formal evaluations** are resourced with dedicated evaluators and are generally scheduled in the long-range or short-range training plans. Formal evaluations are normally highlighted during short-range training briefings. To the maximum extent possible, headquarters two echelons higher conduct formal external evaluations; for example, division commanders evaluate battalions, brigade commanders evaluate companies, and battalion commanders evaluate platoons.
- **Internal evaluations** are planned, resourced, and conducted by the organization undergoing the evaluation.
- **External evaluations** are planned, resourced, and conducted by a headquarters at an echelon higher in the chain of command than the organization undergoing the evaluation or a headquarters outside the chain of command.

6-6. Evaluation of individual and small unit training normally includes every soldier and leader involved in the training. For large-scale training events, evaluators sample a number of individuals and subordinate organizations to determine the likelihood of the entire organization to be able to perform specific mission essential tasks to standard.

6-7. During and after formal evaluation, evaluators prepare their findings and recommendations. They provide these reports to the evaluated unit commander and higher commanders as required by the headquarters directing the evaluation. Evaluation documentation can range from an annotated T&EO for an internal training evaluation to a comprehensive report on RC units during AT periods.

EVALUATION OF TRAINING

6-8. Training evaluations are a critical component of any training assessment. Evaluation measures the demonstrated ability of soldiers, commanders, leaders, battle staffs, and units against the Army standard. Evaluation of training is integral to standards-based training and is the cornerstone of leader training and leader development.

6-9. All training must be evaluated to measure performance levels against the established Army standard. The evaluation can be as fundamental as an informal, internal evaluation performed by the leader conducting the training. Evaluation is conducted specifically to enable the unit or individual undergoing the training to know whether the training standard has been achieved. Commanders must establish a climate that encourages candid and accurate feedback for the purpose of developing leaders and trained units.

6-10. Evaluation of training is not a test; it is not used to find reasons to punish leaders and soldiers. Evaluation tells the unit or the soldier whether or not they achieved the Army standard and, therefore, assists them in determining the overall effectiveness of their training plans. Evaluation produces disciplined soldiers, leaders and units. Training without evaluation is a waste of time and resources.

6-11. Leaders use evaluations as an opportunity to coach and mentor subordinates. A key element in developing leaders is immediate, positive feedback that coaches and leads subordinate leaders to achieve the Army standard. This is a tested and proven path to develop competent, confident adaptive leaders.

AFTER ACTION REVIEW

6-12. The AAR, whether formal or informal, provides feedback for all training. It is a structured review process that allows participating soldiers, leaders, and units to discover for themselves what happened during the training, why it happened, and how it can be done better. The AAR is a professional discussion that requires the active participation of those being trained. The AAR is not a critique and has the following advantages over a critique:

- Focuses directly on key METL derived training objectives.
- Emphasizes meeting Army standards rather than pronouncing judgment of success or failure.
- Uses "leading questions" to encourage participants to self-discover important lessons from the training event.
- Allows a large number of individuals and leaders to participate so more of the training can be recalled and more lessons learned can be shared.

6-13. The AAR consists of four parts—

- **Review what was supposed to happen (training plans).** The evaluator, along with the participants, reviews what was supposed to happen based on the commander's intent for the training event, unit-training plan, training objectives, and applicable T&EOs.
- **Establish what happened.** The evaluator and the participants determine what actually happened during performance of the training task. A factual and indisputable account is vital to the effectiveness of the discussion that follows. For force-on-force training, OPFOR members assist in describing the flow of the training event and discuss training outcomes from their points of view.
- **Determine what was right or wrong with what happened.** The participants establish the strong and weak points of their performance. The evaluator plays a critical role in guiding the discussions so conclusions reached by participants are doctrinally sound, consistent with Army standards, and relevant to the wartime mission.
- **Determine how the task should be done differently the next time.** The evaluator assists the chain of command undergoing the training to lead the group in determining exactly how participants will perform differently the next time the task is performed. This results in organizational and individual motivation to conduct future sustainment training to standard.

6-14. Leaders understand that not all tasks will be performed to standard and in their initial planning, allocate time and other resources for retraining. Retraining allows the participants to apply the lessons learned during the AAR and implement corrective action. Retraining should be conducted at the earliest opportunity to translate observation and evaluation into training to standard. Commanders must ensure that units understand that training is incomplete until the Army standard is achieved.

6-15. The AAR is often "tiered" as a multiechelon leader development technique. Following an AAR with all participants, senior trainers may use the AAR for an extended professional discussion with selected leaders. These discussions usually include a more specific AAR of leader contributions to the observed training results. Commanders use this process as a link between leader training and leader development.

EVALUATORS

6-16. Commanders must plan for formal evaluation and must ensure the evaluators are trained. These evaluators must also be trained as facilitators to conduct AARs that elicit maximum participation from those being trained. External evaluators will be certified in the tasks they are evaluating and normally will not be dual-hatted as a participant in the training being executed. In addition to being able to plan, prepare, and conduct AARs, effective evaluators must also—

- Be familiar with the evaluated organization's METL.
- Be trained (tactically and technically proficient) and rehearsed in the tasks evaluated.
- Know the evaluation standards.
- Follow the tactical and field SOPs for the organization being evaluated.
- Apply relevant information about the evaluated unit, such as wartime missions, personnel turbulence, leader fill, and equipment status.

6-17. Unit leaders, officers and NCOs, must be trained to conduct informal, internal evaluations as well. They must be able to plan, prepare, and conduct AARs effectively. This means they must—

- Be familiar with their organization's METL and how it supports their higher headquarters' METL.
- Be tactically and technically proficient in the tasks evaluated.

6-18. Not only do the individuals and units receiving the training learn from the evaluator, but also the evaluator learns while observing the evaluated unit.

THE ROLE OF SENIOR COMMANDERS AND LEADERS

6-19. Senior commanders ensure that evaluations take place at each echelon in the organization. Commanders use this feedback to teach, coach, and mentor their subordinates. They ensure that every training event is evaluated as part of training execution and that every trainer conducts evaluations. Senior commanders use evaluations to focus command attention by requiring evaluation of specific mission essential and battle tasks. They also take advantage of evaluation information to develop appropriate lessons learned for distribution throughout their commands.

6-20. The use of evaluation data can have a strong effect on the command climate of the organization. Therefore, senior commanders make on-the-spot corrections, underwrite honest mistakes, and create an environment for aggressive action to correct training deficiencies, through retraining.

6-21. Senior commanders use training evaluations as one component of a feedback system. To keep the training system dynamic, they use feedback to determine the effectiveness of the planning, execution, and assessment portions of the training management cycle. These feedback systems allow the senior commander to make changes that lead to superior training results and to teach, coach and mentor subordinate leaders. To be effective, this feedback flows between senior and subordinate headquarters, within each command echelon, and among a network of trainers that may cross several command lines. Some sources of training feedback include—

- Training plan assessments.
- Quarterly training briefing (AC).
- Yearly training briefing (RC).
- PBAC.
- Range conferences.
- Evaluation data.
- Staff visits.
- Leader development discussions.
- Personal observations.
- CTC take home packages.

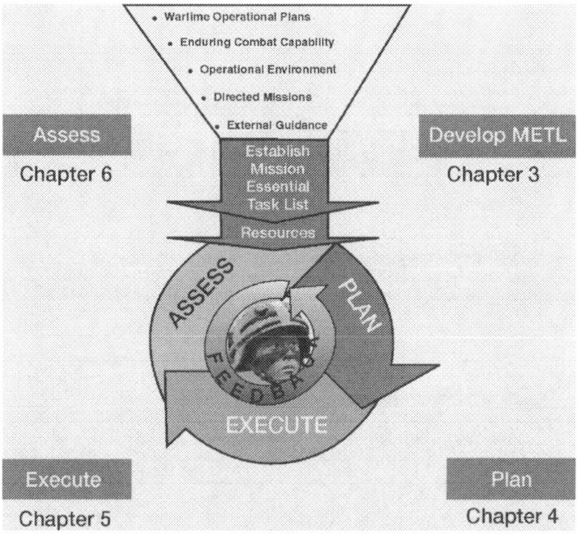

Figure 6-2. Army Training Management Cycle

SUMMARY

6-22. This field manual establishes Army training doctrine and applies throughout the force; to all units, at all echelons, AC and RC. Training to the Army standard is the key to fighting and winning. Every commander and leader from squad through Army is expected to know, understand, and apply this capstone training doctrine. Training excellence is the cornerstone of combat readiness. All leaders are trainers!

The more you sweat in training, the less you bleed in war.

Chinese Proverb

Glossary

SECTION I: ACRONYMS AND ABBREVIATIONS

1SG	first sergeant
A2C2	airspace command and control
AAR	after action review
AC	active component
ACR	armored cavalry regiment
AD	air defense
ADA	air defense artillery
ADC-M	Assistant Division Commander-Maneuver
ADC-S	Assistant Division Commander-Support
ALO	air liaison officer
APFT	Army Physical Fitness Test
APOD	aerial port of debarkation
APOE	aerial port of embarkation
ARFOR	Army forces
ARNG	Army National Guard
ARTEP	Army Training and Evaluation Program
ASOC	air support operations center
ASP	ammunition supply point
AT	annual training
AUTL	Army Universal Task List
AVN	aviation
BASOPS	base operations

BCBST	Brigade Command Battle Staff Training Program (BCBST).
BCTP	Battle Command Training Program
BN	battalion
BOS	battlefield operating system
BRT	brigade reconnaissance troop
C2	command and control
C3I	command, control, communications, and intelligence
C4I	command, control, communications, computers, and intelligence
CALFEX	combined arms live fire exercise
CAS EVAC	casualty evacuation
CBT	combat
CCTT	close combat tactical trainer
CDR	commander
CFX	command field exercise
CHEMO	chemical officer
CMTC	Combat Maneuver Training Center
Co	company
CONUSA	the numbered armies in the continental United States
COSCOM	Corps Support Command
CPC	crew proficiency course
CPX	command post exercise
CS	combat support
CSG	corps support group
CSM	command sergeant major
CSS	combat service support
CTC	combat training center

CTG	command training guidance
CTT	common test training
CTX	combined training exercise
DECOORD	deputy effects coordinator
DEPEX	deployment exercise
DISCOM	division support command
Division (TS)	Division (Training Support)
DIVARTY	division artillery
DMOSQ	duty military occupational specialty qualification
DS	direct support
DS/GS	direct support/general support
EAC	echelons above corps
EAD	echelons above division
ECC	effects coordination cell
ECOORD	effects coordinator
EDRE	emergency deployment readiness exercise
ENCOORD	Engineer Coordinator
ENGR	engineer
EXEVAL	external evaluation
FCX	fire coordination exercises
FRAGO	fragmentary order
FSB	forward support battalion
FSCOORD	Fire Support Coordinator
FSO	fire support officer
FTX	field training exercise

FY	fiscal year
G3	Assistant Chief of Staff for Operations and Plans
GPM	gallons per minute
GSU	Garrison Support Unit
H2O	water
HEMTT	heavy expanded mobility tactical truck
HHC	headquarters and headquarters company
HMMWV	high mobility multipurpose wheeled vehicle
HQ	headquarters
HVY	heavy
IDT	inactive duty training
IFV	infantry fighting vehicle
IG	inspector general
IMT	initial military training
IOM	install, operate, and maintain
IPB	intelligence preparation of the battlefield
ITEP	individual training evaluation program
JFC	joint force commander
JFLC	joint force land component
JFLCC	joint force land component commander
JIM	joint, interagency, multinational
JMETL	joint mission essential task list
JRTC	Joint Readiness Training Center
JTF	joint task force
JTFC	joint task force commander

JTS	Joint Training System
JTX	joint training exercise
LD	line of departure
LFX	live fire exercise
LIN	line item number
LMTV	light medium tactical vehicle
LNO	liaison officer
LOGEX	logistics exercise
LVC	live, virtual, and constructive
M/CM/S	mobility/countermobility/survivability
MACOM	major Army command
MAPEX	map exercise
MCA	movement control agency
MCO	Major Combat Operation
MCOFT	mobile conduct of fire trainer
MDMP	military decision making process
METL	mission essential task list
METT-TC	mission, enemy, terrain, troops, time available, and civil considerations
MI	military intelligence
MILES	multiple integrated laser engagement system
mm	millimeter
MOBEX	mobilization exercise
MOOTW	military operations other than war
MOS	military operational specialty
MP	military police

MRE	mission rehearsal exercise
MSC	major subordinate command
MTA	maneuver training area
MTOE	modification table of organization and equipment
MTP	mission training plan
NBC	nuclear, biological, chemical
NCA	National Command Authority
NCO	noncommissioned officer
NCOES	Noncommissioned Officer Education System
NTC	National Training Center
ODT	overseas deployment for training
OES	Officer Education System
OIP	Organizational Inspection Program
OPFOR	opposing force
OPLAN	operation plan
OPORD	operation order
OPTEMPO	operating tempo
PBAC	Program Budget Advisory Committee
PLT	platoon
PMCS	preventive maintenance checks and services
POL	petroleum, oils and lubricants
PSYOP	psychological operations
QTB	quarterly training brief
QTG	quarterly training guidance
R&S	reconnaissance and security

RC	reserve component
ROE	rules of engagement
ROTC	Reserve Officers' Training Corps
S1	personnel officer
S2	intelligence officer
S3	operations and training officer
S4	logistics officer
S5	civil affairs officer
S6	signal officer
SA	situational awareness
SOF	Special Operations Forces
SOP	standing operating procedure
SPOD	seaport of debarkation
SPOE	seaport of embarkation
STP	Soldier Training Publication
STRAC	Standards in Training Commission
STX	situational training exercise
T&EO	training and evaluation outline
TAA	tactical assembly area
TADSS	training aids, devices, simulators, and simulations
TDA	table of distribution and allowance
TEWT	tactical exercise without troops
TF	task force
TOW	tube-launched, optically tracked, wire-guided
TRADOC	Training and Doctrine Command

TSS	training support system
TTP	tactics, techniques and procedures
TWDS	tactical water distribution system
U.S.	United States
UCOFT	unit conduct of fire trainer
UJTL	Universal Joint Task List
USAF	United States Air Force
USAR	United States Army Reserve
V-C	virtual, constructive
WOES	Warrant Officer Education System
YTB	yearly training brief
YTG	yearly training guidance

SECTION II: TERMS

active component (AC): That portion of the U.S. Army in which organizations are comprised of personnel on full time duty in the active military service of the United States.

after action review (AAR): A method of providing feedback to units by involving participants in the training diagnostic process in order to increase and reinforce learning. The AAR leader guides participants in identifying deficiencies and seeking solutions.

Air Defense Battlefield Operating System: Air defense protects the force from air and missile attack and aerial surveillance. The weapons of mass destruction threat and proliferation of missile technology increase the importance of the air defense system.

annual training (AT): The minimal period of annual active duty training a member performs to satisfy the annual training requirements associated with a reserve component assignment. It may be performed during one consecutive period or in increments of one or more days depending upon mission requirements.

Army Culture: The Army Culture is the Army's shared set of beliefs, values, assumptions about what is important.

Army Service Ethic: The Army Service Ethic is commitment to serve honorably the nation, the Army, its soldiers, and their families above self. This commitment is expressed by the willingness to perform one's duty at all times and to subordinate personal welfare for the welfare of others, without the expectation of reward or recognition. The Army is equally committed to providing values-based leadership and for the well-being of soldiers and their families.

Army Training and Evaluation Program (ARTEP): The cornerstone of unit training. It is the umbrella program to be used by the trainer and training manager in the training evaluation of units. The ARTEP is a complete program enabling commanders to evaluate and develop collective training based on unit weaknesses, then train the unit to overcome those weaknesses and reevaluate. Success on the battlefield depends on the coordinated performance of collective and individual skills that are taught through the ARTEP mission training plan (MTP).

Army Training Management Cycle: The cyclic process of managing and executing training used by Army leaders to identify training requirements and sequentially plan, resource, execute, and evaluate training.

Army Universal Task List (AUTL): The AUTL is a comprehensive listing of Army tactical-level tasks, missions, and operations. The AUTL complements CJCSM 3500.04B, The Universal Joint Task List, by providing tactical-level Army-specific tasks.

Associate AC: Chain of command: The AC/RC Association Program establishes formal linkages between select RC units and an AC MTOE and TDA organization.

band of excellence: The range of proficiency within which a unit is capable of executing its critical wartime tasks, with minimal refresher training, using appropriate repetitions of critical task training.

battlefield operating system (BOS): The physical means used to accomplish the mission. Commanders use BOSs to direct operations. Specifically, commanders arrange BOSs through synchronization to mass effects of combat power at the chosen place or time to overwhelm an enemy or dominate a situation.

battle focus: A concept used to derive peacetime training requirements from assigned and anticipated missions.

battle roster: A listing of individuals, crews, or elements that reflect capabilities, proficiencies in critical tasks, or other information concerning warfighting abilities.

battle task: A task that must be accomplished by a subordinate organization if the next higher organization is to accomplish a mission essential task. The senior commander selects battle tasks from the subordinate organizations' METL.

Brigade Command Battle Staff Training Program (BCBST). This is a Title XI program that provides Enhanced, Divisional, and Strategic Brigades of the Army National Guard the opportunity to sharpen the battle command and battle staff skills. BCBST Program centers on a unit rotation consisting of two major training events: a Battle Command Seminar and a Brigade Warfighter Exercise (BWFX).

Combined Arms Training Strategy (CATS): The Army's overarching strategy for current and future training of the force. It establishes unit, soldier, and leader training requirements and describes how the Army will train and sustain the Army standard in the institution, in units, and through self-development. CATS also identifies and quantifies the training resources required to execute training (AR 350-1).

close combat tactical trainer (CCTT): A virtual simulator trainer that trains tank and mechanized infantry units from platoon to battalion task force, including cavalry scout platoons and heavy cavalry troops on ARTEP MTP collective tasks.

Combat Service Support Battlefield Operating System: Provides the physical means with which forces operate, from the production base and replacement centers in the continental U.S. to soldiers engaged in close combat. CSS includes many technical specialties and functional activities. It includes maximizing the use of host nation infrastructure(s) and contracted support.

Combat Training Center Program: An Army program established to provide realistic joint service and combined arms training in accordance with Army doctrine. It is designed to provide training units opportunities to increase collective proficiency on the most realistic battlefield available during peacetime. The four components of the CTC Program are the—

(1) National Training Center (NTC).

(2) Combat Maneuver Training Center (CMTC).

(3) Joint Readiness Training Center (JRTC).

(4) Battle Command Training Program (BCTP).

combined arms live fire exercises (CALFEX): High-cost, resource intensive exercises in which player units move or maneuver and employ organic and supporting weapon systems using full-service ammunition with attendant integration of all CA, CS, and CSS functions.

combined arms and services training: Collective training that is jointly conducted by associated combat, combat support, and combat service support units.

combined training exercise (CTX): A multinational training event undertaken to enhance U. S. security interests. The exercise is designed to train and evaluate U.S. Forces interoperability with participating Allied nations. The exercise involves planning, preparation, and execution of military maneuvers or simulated wartime and other contingency operations among the United States and other participating Allied nations.

Command and Control (C2) Battlefield Operating System: Command and control has two components—the commander and the C2 system. The C2 system supports the commander's ability to make informed decisions, delegate authority, and synchronize the BOS. Moreover, the C2 system supports commanders' ability to adjust plans for future operations, even while focusing on current operations. Staffs work within the commander's intent to direct units and control resource allocations. Through C2, commanders initiate and integrate all BOS toward a common goal—mission accomplishment.

command field exercise (CFX): A field training exercise with reduced troop and vehicle density, but with full command and control and CSS units.

command post exercise (CPX): An exercise in which the forces are simulated and may be conducted from garrison locations or in between participating headquarters.

command training guidance (CTG): The long-range planning document published by division and brigades (or equivalents) in the active and reserve components to prescribe future training and related activities.

commander/leader assessment: Commanders assessments are subjective in nature and use all available evaluation data and subunit leader input to develop an assessment of the organization's overall capability to accomplish the task. Commanders use the following ratings:

(1) T – Trained. The unit is trained and has demonstrated its proficiency in accomplishing the task to wartime standards.

(2) P – Needs practice. The unit needs to practice the task. Performance has demonstrated that the unit does not achieve the standard without some difficulty or has failed to perform some task steps to standard.

(3) U – Untrained. The unit cannot demonstrate an ability to achieve wartime proficiency.

condition(s): The circumstances and environment in which a task is to be performed.

crawl-walk-run: An objective, incremental, standards-based approach to training. Tasks are initially trained at a very basic level in the crawl stage. Training becomes increasingly difficult in the walk stage. Training approaches the level of realism expected in combat during the run stage.

deployment exercise (DEPEX): An exercise that provides training for individual soldiers, units, and support agencies in the tasks and procedures for deploying from home stations or installations to potential areas of hostilities.

discovery learning: Process that provides opportunity for input and feedback to identify systemic problems and share insights that offer effective solutions.

distributed learning: The delivery of standardized individual, collective, and self-development training to soldiers, civilians, units, and organizations at the right place and time through the use of multiple means and technology. Distributed learning may involve student-instructor interaction in real time and non-real time. It may also involve self-paced student instruction without the benefit of access to an instructor (AR 350-1).

doctrine: Concise expression of how Army forces contribute to unified action in campaigns, major operations, battles and engagements; describes the Army's approach and contributions to full spectrum operations on land; authoritative but requires judgment in its application; rooted in time-tested principles but is adaptable to changing technologies, threats and missions; detailed enough to guide operations, yet flexible enough to allow commanders to exercise initiative within the specific tactical and operational situation; to be useful, doctrine must be well known and commonly understood.

education: Instruction with increased knowledge, skill, and/or experience as the desired outcome for the student. This is in contrast to training, which is based on task performance, and in which specific conditions and standards are used to assess individual and unit proficiency (AR 350-1).

effects coordinator (ECOORD): The field artillery battalion commander serves as the SBCT effects coordinator (ECOORD). He is responsible for all fires and effects planning and coordination for the SBCT. He advises the SBCT commander on the capabilities and employment of fires and effects and is responsible for obtaining the commander's guidance for desired effects and their purpose. The ECOORD is part of the command group and locates where he can best execute the SBCT commander's intent for fires and effects.

engineer coordinator (ENCOORD): The engineer coordinator is the special staff officer for coordinating engineer assets and operations for the command. The ENCOORD is usually the senior engineer officer in the force.

field training exercise (FTX): An exercise conducted under simulated combat conditions in the field. It exercises command and control of all echelons in battle functions against actual or simulated opposing forces.

fire coordination exercise (FCX): An exercise that can be conducted at the platoon, company/team, or battalion/task force level. It exercises command and control skills through the integration of all organic weapon systems, as well as indirect and supporting fires. Weapon densities may be reduced for participating units, and sub-caliber devices substituted for service ammunition.

Fire Support Battlefield Operating System: Fire support consists of fires that directly support land, maritime, amphibious, and special operations forces in engaging enemy forces, combat formations and facilities in pursuit of tactical and operational objectives. Fire support integrates and synchronizes fires and effects to delay, disrupt, of destroy enemy forces, systems, and facilities. The fire support system includes the collective and coordinated use of target acquisition data, indirect fire weapons, fixed-winged aircraft, electronic warfare, and other lethal and non-lethal means to attack targets.

fire support coordinator (FSCOORD): The fire support coordinator is the special staff officer for coordinating fire support and field artillery assets and operations in the command. The FSCOORD is the senior field artillery officer in the force.

force integration: The process of incorporating new doctrine, equipment, and force structure into an organization while simultaneously sustaining the highest possible levels of combat readiness.

inactive duty training (IDT): Authorized training performed by an RC member not on active duty or active duty for training, and consisting of regularly scheduled unit training assemblies, additional training assemblies, or equivalent training periods.

initial military training: Training presented to new enlistees with no prior military service. It is designed to produce disciplined, motivated, physically fit soldiers ready to take their place in the Army in the field. This training consists of BCT, AIT, OSUT, and pre-basic training courses.

Intelligence Battlefield Operating System: A system that plans, directs, collects, processes, produces, and disseminates intelligence on the threat and the environment; performs intelligence preparation of the battlefield (IPB) and other intelligence tasks. Developed as a part of a continuous process and is fundamental to Army operations.

intelligence preparation of the battlefield (IPB): A systematic approach to analyzing the enemy, weather, and terrain in a specific geographic area. It integrates enemy doctrine with the weather and terrain as they relate to the mission and the specific battlefield environment. This is done to determine and evaluate enemy capabilities, vulnerabilities, and probable courses of actions.

interagency coordination: Within the context of Department of Defense (DOD) involvement, the coordination that occurs between elements of DOD, and engaged U.S. Government agencies, nongovernmental organizations, and regional and international organizations for the purpose of accomplishing an objective.

JANUS: An interactive, entity-level, multi-sided, tactical-through-brigade-level simulation used to train junior leaders and to provide battle staff training.

joint mission essential task list (JMETL): A list of joint tasks considered essential to the accomplishment of an assigned or anticipated mission.

leader development: The deliberate, continuous, sequential and progressive process, grounded in Army values, that grows soldiers and civilians into competent and confident leaders capable of decisive action. Leader development is achieved through the life-long synthesis of the knowledge, skills, and experiences gained through the developmental domains of institutional training and education, operational assignments, and self-development.

leader training: Leader training is the expansion of basic soldier skills that qualifies soldiers to lead other soldiers.

leadership: Leadership is influencing people—by providing purpose, direction, and motivation—while operating to accomplish the mission and improving the organization.

learning organization: An organization that is continually expanding its capacity to create its future. It requires a life long commitment to learning and requires all members of the organization, at all levels, to contribute.

life long learning: The individual life long choice to actively and overtly pursue knowledge, the comprehension of ideas, and the expansion of depth in any area in order to progress beyond a known state of development and competency.

live, virtual, constructive: Training environments involving the use of simulations and simulators that provide repetitive, iterative, intense, commander/leader, battle staff, unit and soldier experiences required to achieve and sustain proficiency on critical wartime tasks. The three training environments are—

(1) Live. Training executed in field conditions using tactical equipment, enhanced by training aids, devices, simulators, and simulations (TADSS) and tactical engagement simulation (TES) to simulate combat conditions.

(2) Virtual. Training executed using computer-generated battlefields in simulators with approximate physical layout of tactical weapons systems and vehicles. Virtual TES training permits units to maneuver over much larger areas.

(3) Constructive. The use of computer models and simulations to exercise the command and staff functions of units from platoons through echelons above corps.

logistics exercise (LOGEX): Training exercise that concentrates on training tasks associated with the combat service support battlefield operating system.

map exercise (MAPEX): A training exercise that portrays military situations on maps and overlays that may be supplemented with terrain models and sand tables. It enables commanders to train their staffs in performing essential integrating and control functions under simulated wartime conditions.

mentorship: Mentorship refers to the voluntary developmental relationship that exists between a person of greater experience and a person of lesser experience that is characterized by mutual trust and respect.

military operations other than war (MOOTW): Operations that encompass the use military capabilities across the range of military operations short of war. These military actions can be applied to complement any combination of the other instruments of national power, and occur before, during, and after war.

mission, enemy, terrain and weather, troops and support available, time available and civil considerations (METT-TC): Used to describe the factors that must be considered during the planning, preparation, and execution of full spectrum operations.

mission: The commander's expression of what the unit must accomplish and for what purpose. The primary task assigned to an individual, unit, or force.

mission essential task: A collective task in which an organization must be proficient to accomplish an appropriate portion of its wartime mission(s).

mission essential task list (METL): A compilation of collective mission essential tasks an organization must perform successfully to accomplish its wartime mission(s).

mission rehearsal exercise (MRE): A type of full dress rehearsal that involves every soldier and system participating in the operation and replicates the conditions that the force will encounter during the actual operation; this type of rehearsal produces the most detailed understanding of the mission.

mission training plan (MTP): Descriptive doctrinal training document that provides units a clear description of "what" and "how" to train to achieve wartime mission proficiency. MTPs elaborate on wartime missions in terms of comprehensive training and evaluation outlines, and provide exercise concepts and related training management aids to assist field commanders in the planning and execution of effective unit training.

Mobility/Countermobility/Survivability Battlefield Operating System: Mobility operations preserve the freedom of maneuver for friendly forces. Mobility missions include breaching obstacles, increasing battlefield circulation, improving or building roads, providing bridge and raft support, and identifying routes around contaminated areas. Countermobility denies mobility to enemy forces. Survivability operations protect friendly forces from the effects of enemy weapons systems and from natural occurrences. Nuclear, biological, and chemical defense measures are essential survivability tasks.

multiechelon training: A training technique to train more than one echelon on different tasks simultaneously.

Noncommissioned Officer Education System (NCOES): Prepares noncommissioned officers to lead and train soldiers who work and fight under their supervision and assist their assigned leaders to execute unit missions. NCOES courses provide noncommissioned officers with progressive and sequential leader, technical, and tactical training that is relevant to duties, responsibilities, and missions they will perform in operational units after graduation. Training builds on existing skills, knowledge, behaviors, and experience.

Officer Education System (OES): Produces a corps of broadly-based officer leaders who are fully competent in technical, tactical, and leader skills, knowledge, and behaviors; are knowledgeable of "how the Army runs"; demonstrate confidence, integrity, critical judgment, and responsibility; can operate in an environment of complexity, ambiguity, and rapid change; can build effective teams amid continuous organizational and technological change; and can adapt and solve problems creatively. Officer leader development is a continuous process beginning with pre-commission training and education.

officership: Officership is the practice of being a commissioned Army leader, inspired by a unique professional identity that is shaped by what an officer must Know and Do, but most importantly, by a deeply held personal understanding and acceptance of what an officer must Be. This unique self-concept incorporates our interrelated roles: Warfighter, Servant to the Nation, Member of the time-honored Army Profession, and Leader of Character.

operating tempo (OPTEMPO): The annual operating miles or hours for the major equipment system in a battalion-level or equivalent organization. Commanders use OPTEMPO to forecast and allocate funds for fuel and repair parts for training events and programs.

organizational assessment: A process used by Army senior leaders to analyze and correlate evaluations of various functional systems, such as training, logistics, personnel, and force integration to determine an organization's capability to accomplish its wartime mission.

Profession of Arms: The fundamental characteristics of Army professionalism are a service focus, an expert knowledge, a unique culture, and a professional military ethos. Army professionalism is intellectual, physical, and moral in nature; intellectual because of the unique and extensive body of expertise required in military operations; physical because of the physical demands of the application of force and the requirement to communicate this real capability to an adversary; moral because the capability to wield tools of destruction in a brutal environment carries with it a moral responsibility.

Program Budget Advisory Committee (PBAC): A committee comprised of the principal staff officers of a command, agency, or installation headquarters, and established for the purpose of coordinating program and budget actions within the command.

pre-execution checks: The informal planning and detailed coordination conducted during preparation for training.

quarterly training brief (QTB): A conference conducted by AC division commanders to approve the short-range plans of battalion commanders.

quarterly training guidance (QTG): An active component training management document published at each level from battalion to division that addresses a three-month planning period. The QTG adjusts, as required, and further develops the training guidance contained in long-range plans, to include specific training objectives for each major training event.

risk management: The process of identifying, assessing, and controlling risks arising from operational factors and making decisions that balance risk costs with mission training benefits.

reserve component (RC): Individuals and units assigned to the Army National Guard or the U.S. Army Reserve, who are not in active service, but who are subject to call to active duty.

round out: RC units that are designated to fill the organizational structure of AC divisions.

self-development: A self-directed, competency-based, progressive, life-long process soldiers use to augment institutional training and unit experience to attain proficiency at their current rank/assignment, and to prepare for promotion and higher-level responsibilities. Self-development is an individual responsibility, assisted by first line leaders and commanders, to identify requirements based on self-assessment and feedback. Development activities are planned to meet specific individual training goals and needs.

situational training exercise (STX): A mission-related, limited exercise designed to train one collective task, or a group of related tasks or drills, through practice.

standard: The minimum acceptable proficiency required in the performance of a particular training task under a specified set of conditions.

Standards in Training Commission (STRAC): Provides coordination and synchronization of resources for CATS. Issues between CATS and STRAC resourcing of strategies are resolved through the Training and Leader Development General Officer Steering Committee (TLGOSC) process.

tactical exercise without troops (TEWT): An exercise conducted in the field on actual terrain suitable for training units for specific missions. It is used to train subordinate leaders and battle staffs on terrain analysis, unit and weapons emplacement, and planning the execution of the unit mission.

training aids, devices, simulators, and simulations (TADSS): A general term that includes combat training centers and training range instrumentation; tactical engagement simulation (TES); battle simulations; targetry; training-unique ammunition; dummy, drill, and inert munitions; casualty assessment systems; graphic training aids; and other training support devices.

task: A clearly defined and measurable activity accomplished by individuals and organizations. Tasks are specific activities that contribute to the accomplishment of encompassing missions or other requirements.

task organization: A temporary grouping of forces designed to accomplish a particular mission.

The Army School System: The fully accredited and integrated Active Army, Army National Guard, and U.S. Army Reserve schools that provide standard resident and nonresident (distance learning) training and education for the Army.

training: The instruction of personnel to increase their capacity to perform specific military functions and associated individual and collective tasks.

training and evaluation outline (T&EO): A summary document prepared for each training activity that provides information on collective training objectives, related individual training objectives, resource requirements, and applicable evaluation procedures.

training assessment: This is a commander's responsibility. It is the commander's judgment of the organization's ability to accomplish its wartime mission. An analytical process used by the Army. The commander bases the training assessment on an analysis of evaluations and other sources of feedback to determine an organization's current levels of training proficiency on mission essential tasks.

training evaluation: The process used to measure the demonstrated ability of individuals and units to accomplish specified training objectives.

training management: The process used by Army leaders to identify training requirements and to subsequently plan, resource, execute, and evaluate training.

training meeting: A periodic meeting conducted by platoon, company, and battalion key leaders to review past training, plan and prepare future training, and exchange timely training information between participants.

training objective: A statement that describes the desired outcome of a training activity. A training objective consists of the following three parts:

(1) Task. A clearly defined and measurable activity accomplished by individuals or organizations.

(2) Condition(s). Describes the circumstances and environment in which a task is to be performed.

(3) Standard. The minimum acceptable proficiency required in the performance of a particular training task.

training requirements: The difference between demonstrated performance and the Army standard of proficiency for mission essential or battle tasks.

training resources: Those resources (human, physical, financial, and time) used to support training. They may be internally controlled by an organization or externally controlled by a headquarters that allocates their use to units as required.

training schedule: A document prepared at company level that specifies the 'who, what, when, and where' of training to be conducted by the unit.

training strategy: The method(s) used to attain the Army standard of training proficiency on mission essential tasks.

Training Support System (TSS): A system of systems that include information technologies; training aids, devices, simulations, and simulators (TADSS); and training support products, services, and facilities. These components are linked by architectures and standards that enable their interconnectivity and interoperability to ensure operationally relevant training experiences for warfighters. The TSS employs management, evaluation, and resource processes to ensure the entire system is assessed, funded, and managed for optimum benefit.

Universal Joint Task List (UJTL): A structured listing of tasks that describe the functional capabilities that joint force commanders may require to execute their assigned missions.

Warrant Officer Education System (WOES): Develops a corps of highly specialized experts and trainers who are fully competent in technical, tactical, and leader skills, knowledge, and behaviors; who are creative problem solvers able to function in highly complex and dynamic environments; and who are proficient operators, maintainers, administrators, and managers of the Army's equipment, support activities, and technical systems. Warrant officer leader development is a continuous process beginning with pre-appointment training and education.

Warrior Ethos: Warrior Ethos compels soldiers to fight through all conditions to victory no matter how much effort is required. It is the soldier's selfless commitment to the nation, mission, unit, and fellow soldiers. It is the professional attitude that inspires every American soldier. Warrior Ethos is grounded in refusal to accept failure. It is developed and sustained through discipline, commitment to the Army values, and pride in the Army's heritage.

well-being: Well-being is the personal, physical, material, mental, and spiritual state of soldiers, civilians, and their families that contributes to their preparedness to perform the Army's mission.

yearly training brief (YTB): A conference conducted by reserve component division commanders to approve the short-range plans of battalion commanders.

yearly training guidance (YTG): A reserve component training management document published at each level from battalion to division that addresses a one-year planning period. The YTG adjusts, as required, and further develops the training guidance contained in long-range plans, to include specific training objectives for each major training event.

References

SECTION I: REQUIRED PUBLICATIONS

FM 1, *The Army*, 14 June 2001

FM 3-0, *Operations*, 14 June 2001

FM 22-100, *Army Leadership*, 31 August 1999 (To be replaced by FM 6-22)

FM 100-14, *Risk Management*, 23 April 1998

FM 101-5, *Staff Organization and Operations*, 31 May 1997 (To be replaced by FM 5-0 and FM 6-0)

FM 101-5-1, *Operational Terms and Graphics*, 30 September 1997 (To be renumbered FM 1-02)

SECTION II: RELATED PUBLICATIONS

AR 1-201, *Army Inspection Policy*, 17 May 1993

AR 20-1, *Inspector General Activities and Procedures*, 29 March 2002

AR 220-1, *Unit Status Reporting*, 30 November 2001

AR 350-1, *Army Training*, 1 August 1981

AR 350-41, *Training in Units*, 19 March 1993

AR 350-50, *Combat Training Center Program*, 24 May 1995

AR 385-16, *System Safety Engineering and Management*, 2 November 2001

AR 600-20, *Army Command Policy*, 13 May 2002

ARTEP 5-415-66-MTP, *Mission Training Plan for the Engineer Battalion (Combat) (Heavy), Battalion Staff,* 26 July 2002

ARTEP 5-417-17-MTP, *Mission Training Plan for the General Construction Platoon, Engineer Company (Combat) (Heavy),* 26 July 2002

ARTEP 5-417-35-MTO, *Mission Training Plan for the Engineer Company, Engineer Battalion (Combat) (Heavy),* 26 July 2002

ARTEP 7-7J-DRILL, *Battle Drills for the Bradley Fighting Vehicle Platoon, Section and Squad,* 8 December 1992

ARTEP 10-468-30-*MTP, Mission Training for the Quartermaster Company (Water Supply) (DS/GS) and Quartermaster Tactical Water Distribution Team (Hoseline),* 1 April 2002

ARTEP 71-1-MTP, *Mission Training Plan for Tank and Mechanized Infantry Company and Company Team,* 28 April 1994

ARTEP 71-2-MTP, *Mission Training Plan for the Tank and Mechanized Infantry Battalion Task Force,* 27 November 2001 (EOM)

CJCSI 3500.01B, *Joint Training Policy for the Armed Forces of the United States*, 31 December 1999

CJCSM 3500.03, *Joint Training Manual for the Armed Forces of the United States*, 1 June 1996

CJCSM 3500.04C, *Universal Joint Task List (UJTL),* 1 July 2002

DA Pam 350-38, *Standards in Weapons Training,* 3 July 1997

DA Pam 350-39, *Standards in Weapons Training (Special Operations Forces),* 3 July 1997

DA Pam 350-58, *Leader Development for America's Army,* 13 October 1994

FM 25-101, *Battle Focused Training,* 30 September 1990

FM 100-6, *Information Operations,* 27 August 1996 (To be replaced by FM 3-13)

FM 100-8, *The Army in Multinational Operations,* 24 November 1997 (To be replaced by FM 3-16 Multinational Operations)

FM 100-19, *Domestic Support Operations,* 1 July 1993 (To be replaced by FM 3-07.7 Stability Operations and Support Operations)

FORSCOM/ARNG/USAR Reg 350-2, Reserve Component Training, 27 October 1999

JP 1-02, *DOD Dictionary of Military and Associated Terms,* 12 April 2001

JP 3-07, *Joint Doctrine for Military Operations Other Than War,* 16 June 1995

JP 3-08, *Interagency Coordination During Joint Operations,* 9 October 1996

JP 3-16, *Joint Doctrine for Multinational Operations,* 5 April 2000

JP 3-33, *Joint Force Capabilities,* 13 October 1999

STP 10-77W14-SM-TG, *Soldier's Manual and Trainer's Guide for Water Treatment Specialist MOS 77W – Skill Levels 1, 2, 3 and 4,* 12 November 1993

TRADOC Reg 350-10, *Institutional Leader Training and Education,* 12 August 2002

TRADOC Reg 350-70, *Systems Approach to Training Management, Processes and Products,* 9 March 1999

TRADOC Pam 350-70-1, *Guide for Producing Collective Training Products,* 15 March 1996

"Risk Management for Brigades and Battalions", Center for Army Lessons Learned (CALL) Newsletter 99-5, 9 June 1995

SECTION III: PROJECTED PUBLICATIONS

FM 5-0 Army Planning and Orders Production.

FM 6-0, *Command and Control*

FM 7-1 Battle Focused Training

FM 7-15, *The Army Universal Task List (AUTL)*

Index

A

active component, 1-13, 3-6, 4-5 to 4-9, 4-17 to 4-31, 5-3, 6-6

after action review, 1-11, 2-2, 5-3, 5-6, 5-7, 6-1, 6-2, 6-3, 6-4

annual training, 4-9, 4-12, 4-19, 4-20, 4-27, 6-2, 6-3

Army Training and Evaluation Program, 6-2

Army Training Management Cycle, 2-1, 2-14, 2-15

Army Universal Task List, 3-4, 3-13

associate AC, 3-6

B

Band of Excellence, 2-9, 3-11, 4-3, 4-9, 4-13, 4-15, 4-20, 4-21, 4-22, 5-4,

battlefield operating system, 2-3 to 2-5, 3-9, 3-10, 4-14, 5-6

battle focus, 2-1, 2-8, 2-13, 2-14, 3-5, 3-6, 3-8, 3-15, 4-1, 4-5, 4-28

battle task, 2-8, 3-14, 3-15, 4-5, 4-8, 4-14, 4-15, 4-17, 4-28, 6-5

battle roster, 5-5, 5-6

C

close combat tactical trainer, 4-20, 4-27

command training guidance, 4-7, 4-8, 4-9, 4-22

combined arms live fire exercise, 4-14

combined training exercise, 4-14

command and control, 1-4, 3-5, 3-7, 3-9, 3-10, 3-14, 4-3, 5-5

command field exercise, 4-14, 4-20

command post exercise, 4-3, 4-14, 4-20, 4-27

command training guidance, 4-7, 4-8, 4-9, 4-21

conditions, iv, 1-2, 1-5, 1-13, 2-1, 2-6, 2-7, 3-8, 3-9, 3-10, 3-12, 3-13, 4-6, 4-14, 4-16, 4-20, 4-28, 4-33, 5-3, 5-4, 5-5

crawl-walk-run, 5-3, 5-4

D

deployment exercise, 1-10, 3-7, 4-14, 4-20

distributed learning, 1-8, 1-9, 1-12, 4-10

doctrine, iv, 2-7, 1-3, 1-4, 1-5, 1-7, 1-9, 1-11, 1-13, 1-14, 2-3, 2-7, 2-10, 3-8, 4-4, 4-29, 5-4, 5-6, 6-6

E

education, 1-5 to 1-8, 1-10, 1-13, 4-10, 4-11, 4-13, 4-29

F

field training exercise, 4-14, 4-18, 4-20, 4-21, 4-25, 4-26, 4-33, 5-6, 6-2

fire coordination exercise, 4-14

force integration, 3-4, 4-4, 4-8, 4-17, 4-23, 4-29, 6-1, 6-2

I

initial military training, 1-8

J

JANUS, 4-20

joint mission essential task list, 3-8

L

leader development, iv, v, 1-5, 1-6, 1-7, 1-10, 1-11 to 1-14, 2-10, 2-12, 3-2, 4-23, 4-29, 4-30, 5-6, 6-3 to 6-5

leader training, 1-6 to 1-9, 2-10, 3-7, 4-8, 4-16, 4-23, 4-28, 5-2, 5-5, 6-3, 6-4

leadership, iv, 1-6 to 1-9, 1-11, 2-11, 5-7

logistics exercise, 4-14

M

map exercise, 4-14

military operations other than war, 1-2, 1-3

mission essential task list, 1-8, 1-9, 1-10, 1-14, 2-1, 2-5, 2-6, 2-8, 2-9, 2-14, 2-15, 3-2 to 3-9, 3-10, 3-11, 3-13, 3-14, 4-1 to 4-5, 4-8, 4-9, 4-10, 4-13 to 4-17, 4-20, 4-23, 4-24, 4-25, 4-27, 4-29, 4-30, 4-32, 4-33, 5-2, 5-4, 5-6, 5-8, 6-3, 6-4, 6-5

mission rehearsal exercise, 3-7

mission training plan, 1-9, 1-10, 2-7, 2-8, 3-4, 3-11, 3-

12, 4-2, 4-26, 4-27, 4-33, 4-34, 5-4, 5-5

mission, enemy, terrain and weather, troops and support available, time available and civil considerations, 5-5

multiechelon training, 2-8, 4-11, 4-13, 4-14, 4-24, 4-25, 4-26, 4-27

N

Noncommissioned Officer Education System, 1-8

O

Officer Education System, 1-8

operating tempo, 4-19

organizational assessment, 2-15, 6-1, 6-2

P

pre-execution checks, 4-31, 5-2, 5-3

program budget advisory committee, 4-27, 5-4, 6-5

Q

quarterly training brief, 4-22, 4-28, 4-29, 6-5

quarterly training guidance, 4-22, 4-23, 4-28, 4-29

R

reserve component, 1-12, 1-13, 2-9, 2-10, 3-2, 3-5, 3-6, 4-5, 4-6, 4-8, 4-9, 4-12, 4-21, 4-22, 4-26, 4-27, 4-28, 4-30, 4-31, 5-3, 6-3, 6-5, 6-6

risk management, 3-12, 4-6, 4-9, 4-23, 4-29, 5-2, 5-4

S

self-development, 1-5 to 1-7, 1-11, 1-12, 1-14, 2-12, 2-13,

situational training exercise, 1-10, 4-14, 4-20, 4-25, 4-29, 4-34, 5-3

standard, iv, 1-1, 1-2, 1-4, 1-5, 1-8, 1-9, 1-13, 2-1, 2-2, 2-4, 2-6, 2-7, 2-11 to 2-14, 3-7, 3-11, 3-12, 3-13, 4-2, 4-3, 4-6, 4-9, 4-13, 4-15, 4-16, 4-21, 4-24, 4-27 to 4-29, 4-30, 4-32, 4-33, 5-2, 5-3, 5-4, 5-5, 5-7, 6-1, 6-3, 6-4, 6-6

T

tactical exercise without troops, 4-3, 4-14, 4-20

task, iv, 1-2 to 1-5, 1-7 to 1-10, 1-12, 1-13, 2-2 to 2-5, 2-8, 2-13, 2-14, 3-4 to 3-12, 4-2 to 4-6, 4-8, 4-10 to 4-14, 4-16, 4-17, 4-21, 4-23, to 4-30, 4-33, 4-34, 5-1, 5-2, 5-3, 5-4, 6-4, 6-5

task organization, 2-3, 3-10, 4-5, 4-30

training aids, devices, simulators and simulations, 2-7, 4-16, 4-17, 4-28, 5-4, 5-6

training and evaluation outline, 4-14, 5-3, 5-5, 6-4

training assessment, 2-15, 4-2 to 4-4, 6-1, 6-4

training evaluation, 2-16, 4-2, 4-8, 4-23, 6-1, 6-4

training management, iv, v, 1-12, 2-1, 4-8, 4-23, 4-31

training meeting, 4-5, 4-30, 4-31, 5-2, 5-8

training objective, 3-11, 3-12, 4-5, 4-13, 4-23, 4-30, 5-2, 5-4, 5-5, 5-6, 6-4

training requirements, 1-13, 2-11, 2-13, 3-5, 3-8, 4-14, 4-17, 4-22, 5-1

training resources, 3-11, 4-17, 4-20, 4-27, 5-2, 5-4, 5-7

training schedule, 4-5, 4-23, 4-30, 4-31, 4-32, 5-2, 5-8

training strategy, 2-7, 2-9, 3-10, 4-3, 4-21

Training Support System, 1-13, 4-5, 4-14, 4-17, 4-27, 5-3, 5-4,

U

Universal Joint Task List, 3-8

W

Warrant Officer Education System, 1-8

warrior ethos, 1-5, 1-7, 1-8

Y

yearly training brief, 4-22, 4-23, 4-28, 4-29

yearly training guidance, 4-22, 4-23, 4-28, 4-29

Made in United States
Troutdale, OR
11/03/2026

56844378R00066